Lecture Notes in Mathematics

Edited by A. Dold and B. Eckmann

1268

S. G. Krantz (Ed.)

T0222446

Complex Analysis

Seminar, University Park PA, March 10–14, 1986

Springer-Verlag
Berlin Heidelberg New York London Paris Tokyo

Editor

Steven G. Krantz
Department of Mathematics, Washington University
St. Louis, Missouri 63130, USA

Mathematics Subject Classification (1980): 32 A 17, 32 A 10, 32 B 10, 32 H 15, 32 M 10, 32 F 05, 32 F 15

ISBN 3-540-18094-X Springer-Verlag Berlin Heidelberg New York
ISBN 0-387-18094-X Springer-Verlag New York Berlin Heidelberg

© Springer-Verlag Berlin Heidelberg 1987
Printed in Germany

Printing and binding: Druckhaus Beltz, Hemsbach/Bergstr.
2146/3140-543210

PREFACE

This volume represents the proceedings of an intensive week of complex analysis at Penn State which was held during the week of March 10, 1986. The conference was attended by about fifteen people with similar interests, and every participant attended every lecture. The result was an enjoyable and rewarding exchange of ideas.

The lead article in this volume is a rather personal assessment of progress in Several Complex Variables in the past fifteen years. Subsequent articles in the volume point to a number of new paths which we expect the subject to follow. We hope that the volume will be especially helpful to students and new members in the field, as well as to people who are already established.

We are grateful to the Department of Mathematics and the College of Science at the Pennsylvania State University for funding this conference.

Steven G. Krantz
St. Louis, Missouri USA
March, 1987

CONFERENCE PARTICIPANTS

David E. Barrett	Princeton University
Eric Bedford	Indiana University
Jay Belanger	Princeton University
Steven R. Bell	Purdue University
John Bland	Tulane University
Joseph A. Cima	University of North Carolina
John P. D'Angelo	University of Illinois
John Erik Fornaess	Princeton University
K. T. Hahn	Pennsylvania State University
Steven G. Krantz	Pennsylvania State University
Donald Rung	Pennsylvania State University
Rita Saerens	Michigan State University
Berit Stensønes	Rutgers University

TABLE OF CONTENTS

Steven G. Krantz, Recent Progress and Future Directions in Several Complex Variables 1

David E. Barrett, Boundary Singularities of Biholomorphic Maps 24

S. Bell, Compactness of Families of Holomorphic Mappings up to the Boundary 29

J. S. Bland, The Imbedding Problem for Open Complex Manifolds 43

J. S. Bland, T. Duchamp and M. Kalka, A Characterization of CP^n by its Automorphism Group 60

Joseph A. Cima and Ted Suffridge, Proper Mappings Between Balls in C^n 66

John P. D'Angelo, Finite-Type Conditions for Real Hypersurfaces in C^n 83

John P. D'Angelo, Iterated Commutators and Derivatives of the Levi Form 103

John Eric Fornaess and Nessim Sibony, Plurisubharmonic Functions on Ring Domains 111

Robert E. Greene and Steven G. Krantz, Characterizations of Certain Weakly Pseudoconvex Domains with Non-Compact Automorphism Groups 121

Rita Saerens, Interpolation Theory in C^n: A Survey 158

Berit Stensones, Extendability of Holomorphic Functions 189

Recent Progress and Future Directions in Several Complex Variables

Steven G. Krantz
Department of Mathematics
Washington University
St. Louis, Missouri 63130

Section 0 Introduction

I doubt that anyone is qualified to produce a list of the most important results in several complex variables from the last ten or fifteen years--certainly I am not. What I want to do here is to comment on a few areas with which I have some familiarity, and on which the talks given at this conference impinge. A lot has been written about the fallout from the construction of integral representations by Henkin, Ramirez, and others (see [HEN1] and [RAM]). And the significance and pervasiveness of the theory of the $\bar{\partial}$-Neumann problem and the weighted L^2 estimates of Hörmander seem to be well-known. Therefore I would rather concentrate here on the consequences of a few key events which occurred in the early 1970's. These are the following:

(i) The Kohn-Nirenberg example (1973)
(ii) The worm domain (Diederich and Fornaess, 1977)
(iii) Bounded strictly plurisubharmonic exhaustion functions (Diederich and Fornaess, 1977)
(iv) Points of finite type and subelliptic estimates for the $\bar{\partial}$-Neumann problem (Kohn, 1972)
(v) Fefferman's mapping theorem (1974)

It is my view that these discoveries, and related results by dozens of other mathematicians, have completely altered the way that we think about several complex variables and, in particular, they have changed the course of research.

The fundamental change that has occurred is simply this: much of the work prior to the early 1970's, and even the work that goes into the original proof of Fefferman's mapping theorem, entailed a very careful understanding of <u>strongly</u> pseudoconvex points. The Levi problem had been reduced to the strongly pseudoconvex case in the 1930's, and the solution of the Levi problem in the 1950's included a detailed analysis of the local geometry of the positive definite Levi form (see, for instance, [BER]). Likewise, Kohn's solution of the $\bar{\partial}$-Neumann problem (see [FOK]) hinged on a consideration of the Neumann boundary conditions at strongly pseudoconvex points. Fefferman's work in [FE1] required a much deeper analysis of the approximations of strongly pseudoconvex domains by balls. It is safe to say that Fefferman's work in [FE1], [FE2], [FE3], including work of Folland/Stein [FOS1] and others that it built on, represents the deepest understanding to date of strong pseudoconvexity.

What is very striking is that, until the early 1970's, no one had thought carefully about weakly pseudoconvex points. As recently as 1972, experts were still conjecturing that a weakly pseudoconvex point is, up to local biholomorphic equivalence, weakly convex (this would be in analogy with the fact, which in 1972 had been known for nearly twenty years, that a strongly pseudoconvex point is locally biholomorphically equivalent to a strongly convex point). We now realize that such a conjecture is not even approximately correct, at least not in a form that gives useful results. In fact we are only beginning to understand weakly pseudoconvex points.

By the same token, it has long been known that weakly pseudoconvex domains can be exhausted by smoothly bounded strongly pseudoconvex domains. It was generally believed that weakly pseudoconvex domains could be approximated from the outside by strongly pseudoconvex domains. That such is not the case (see [DF1]) came as quite a shock. See Section 1 for further details.

In the next three sections I would like to use items (i) - (v) listed above as a vehicle for discussing what we have learned about pseudoconvexity, and what remains to be done. It should be stressed that I am concentrating on only a portion of the theory of several complex variables--that which derives primarily from the Princeton school. But it is this collection of ideas with which I am most familiar, and which to my mind has yielded the most novel ideas and techniques in the past decade or so. While the inner functions problem, the corona problem, and many other programs have also had strong effects on several complex variables, they are not salient to the theme of this paper and I shall not discuss them.

The reader is advised to refer to [HO] or [KR1] or [RAN2] for basic definitions which bear on the discussion that follows.

<u>Section 1</u> The Kohn-Nirenberg Example, the Worm Domain, and Related Phenomena

The most important elementary fact about a strongly pseudoconvex point P in the boundary of a domain Ω is the existence of a <u>local</u>

<u>holomorphic</u> <u>separating</u> <u>function</u> for Ω at P. Indeed, if

$$\Omega = \{z \in \mathbb{C}^n : \rho(z) < 0\}$$

satisfies

$$\sum_{j,k=1}^{n} \frac{\partial^2 \rho}{\partial z_j \partial \bar{z}_k}(P) w_j \bar{w}_k \geq C|w|^2 \qquad \forall w \in \mathbb{C}^n$$

then

$$L_P(z) \equiv \sum_{j=1}^{n} \frac{\partial \rho}{\partial z_j}(P)(z_j - P_j) + \sum_{j,k=1}^{n} \frac{\partial^2 \rho}{\partial z_j \partial z_k}(P)(z_j - P_j)(z_k - P_k)$$

satisfies

$$\bar{\Omega} \cap \{z : |z - P| < \epsilon_0 \, , \, L_P(z) = 0\} = \{P\}$$

when $\epsilon_0 > 0$ is small. Alternatively, once one notices that there is a local biholomorphic change of coordinates near P which renders $\partial\Omega$ strongly <u>convex</u>, say that in the new coordinates $(w_1, \ldots, w_n)$ the domain Ω near P has defining function $\rho^*(w)$ and that $P \longleftrightarrow P^*$, then it is clear that the pullback of

$$\varphi(w) \equiv \Sigma \frac{\partial \rho^*(P^*)}{\partial w_j} \cdot (w_j - P^*_j)$$

near P will be a local holomorphic separating function.

The existence of holomorphic separating functions is a critical step in the solution of the Levi problem (see [BER]). In the construction of integral formulas using the Cauchy-Fantappiè machinery, the existence of <u>global</u> holomorphic separating functions (gotten from local holomorphic separating functions by solving a suitable cohomology problem) is fundamental. Holomorphic separating functions provide important information about optimal regularity for the $\bar{\partial}$ problem (see [KR2]). Finally, holomorphic separating functions are very closely related to holomorphic peaking functions which, in turn, are basic for function algebraic considerations.

Were the aforementioned conjecture, that smoothly bounded weakly pseudoconvex domains are locally biholomorphically equivalent to weakly convex domains, in fact true then the pullback of

$$\varphi(z) \equiv \Sigma \frac{\partial \rho^*(P^*)}{\partial w_j} \cdot (w_j - P^*_j) \ ,$$

where $\rho^*(w)$ is a defining function for the convex domain, would give a <u>weak</u> local holomorphic separating function h_P at each point of the boundary. This would mean that

$$P \in \bar{\Omega} \cap \{z : |z - P| < \epsilon_0 , h_P(z) = 0\} \subseteq \partial\Omega .$$

In 1973 Kohn and Nirenberg [KON] destroyed this optimistic program by proving that the point (0,0) in the boundary of the smooth, pseudoconvex domain

$$\Omega = \{(z_1,z_2) \in \mathbb{C}^2 : \operatorname{Re} z_2 + |z_1 z_2|^2 + |z_1|^8 + \frac{15}{7}|z_1|^2 \operatorname{Re}(z_1)^6 < 0\}$$

has no local holomorphic separating function. Indeed if h is holomorphic in a neighborhood of 0 and $h(0) = 0$ then h vanishes infinitely often <u>on</u> <u>Ω</u> in every neighborhood of 0 . In particular we see that the Kohn-Nirenberg domain is not locally biholomorphically equivalent to a convex domain near 0 . Hakim/Sibony and Sibony (see [HS1],[SI1]) have subsequently obtained stronger examples which show that weakly pseudoconvex boundaries cannot necessarily be made locally convex even with a biholomorphism from one side.

We now understand that the domains constructed by Kohn/Nirenberg and Hakim/Sibony are weakly pseudoconvex domains of the most tractable sort: that is, the the Levi form vanishes at the bad points, but only to finite order. Such points are called finite type (see Section 2) and for many purposes they are as good as strongly pseudoconvex points. The lesson to be learned is that even if one restricts attention to the simplest weakly pseudoconvex domains, even in $\mathbb{C}^2$, the notion that pseudoconvexity is a biholomorphically invariant version of convexity is far too simple-minded.

The Kohn-Nirenberg example inspired a number of people to investigate holomorphic separating functions, peak points, Šilov boundaries, and related phenomena. The papers [HS1], [BL1], [BL2], [SI1], [SI2], [FO1], [BEF] give an overview of some of this work.

Another drive to reduce the study of weakly pseudoconvex domains to the more tractable strongly pseudoconvex domains was the problem of the Nebenhülle. If Ω is a (pseudoconvex) domain then Ω is said to have a <u>Stein</u> <u>neighborhood</u> <u>basis</u> if $\Omega = \cap \Omega_j$, each Ω_j is strongly pseudoconvex, and $\Omega_j \supseteq \Omega_{j+1}$, each j . If Ω does not have a Stein neighborhood basis then

$$\cap \{\Omega' : \Omega' \supseteq \Omega , \Omega' \text{ is strongly pseudoconvex}\}$$

is called the <u>Nebenhülle</u> of Ω .

It was commonly supposed, if not fervently hoped, that every smooth pseudoconvex domain has a Stein neighborhood basis. In retrospect, this was probably a bit optimistic. For the Hartogs triangle

$$T = \{(z_1,z_2) : |z_1| < |z_2| < 1\}$$

has a very large Nebenhülle. To be sure, ∂T is only Lipschitz, but there is no substantive reason why smoothly bounded domains should be better behaved.

In any event, it was quite a surprise when in 1977 Diederich and Fornaess [DF1] exhibited the "worm domain": a smoothly bounded pseudoconvex domain with non-trivial Nebenhülle. It was a difficult lesson to accept that pseudoconvex domains look a lot different from the inside than from the outside--in particular they are much more subtle than convex domains. But this remarkable discovery gave a great impetus to the research of the 1970's.

I would be remiss at this point not to mention the one piece of good news that came along in this time period: the discovery of the bounded strictly plurisubharmonic exhaustion functions [DF2]. In fact, given any smoothly bounded pseudoconvex domain Ω then there is a defining function ρ for Ω and an $\eta > 0$ such that $\hat{\rho} \equiv -(-\rho)^{\eta}$ satisfies

(i) $\hat{\rho}$ is strictly plurisubharmonic on Ω ;
(ii) $\hat{\rho} < 0$ on Ω , $\hat{\rho} = 0$ on $\partial\Omega$;
(iii) $\Omega_c \equiv \{z \in \Omega : \hat{\rho} \leq c\} \subset\subset \Omega$, all $c < 0$;
(iv) If $K \subset\subset \Omega$ then there is a $c < 0$ such that $K \subseteq \Omega_c$.

For many purposes, the bounded plurisubharmonic exhaustion function is a good substitute for the program that the Kohn-Nirenberg example and the Diederich-Fornaess example killed. It has proved particularly useful in studying the holomorphic mapping problem (see for instance [BEL] and [DF4]).

One very important lesson that was learned from the Kohn-Nirenberg example and the two results of Diederich and Fornaess is that there is no substitute in several complex variables for hard calculations. These papers set the tone for the decade of research that followed.

<u>Section 2</u> Points of Finite Type and Subelliptic Estimates

Let $\Omega = \{z : \rho(z) < 0\} \subseteq \mathbb{C}^2$ have smooth boundary. A point $P \in \partial\Omega$ is said to be of <u>finite type</u> $m \in \mathbb{Z}^+$ if there is a nonsingular complex variety V such that

$$|\rho(v)| \leq C|v - P|^{m+1} \quad , \quad v \in V$$

while there is no nonsingular complex variety V' such that

$$|\rho(v')| \leq C|v' - P|^{m+2} \quad , \quad v' \in V' .$$

The notion of finite type is unoriented: it cannot distinguish between pseudoconvexity and pseudoconcavity. Thus it turns out that the only points of type 1 are strongly pseudoconvex points and strongly pseudoconcave points. Pseudoconvex points are always of odd type. In the domain

$$\{(z_1,z_2) : |z_1|^2 + |z_2|^{2k} < 1\},$$

boundary points of the form $(e^{i\theta},0)$ are of type $2k - 1$.

The notion of finite type helps us to quantify the idea that strongly pseudoconvex points are generic in the boundaries of smooth pseudoconvex domains. For if $U \subseteq \partial\Omega \subseteq \mathbb{C}^2$ is a relatively open subset containing only points of finite type exceeding one, then U consists only of points where the Levi form vanishes. In other words, U consists only of points where the Levi form has zero rank. It follows (see [KR1]) that U is foliated by one dimensional complex manifolds. As a result, each point of U is of infinite type, and that is a contradiction.

Continuing to restrict attention to $\mathbb{C}^2$, we now give another (equivalent) definition of finite type. If $\Omega = \{z: \rho(z) < 0\}$ is a smoothly bounded domain in $\mathbb{C}^2$, $P \in \partial\Omega$, and $\frac{\partial\rho}{\partial z_2}(P) \neq 0$, we define a vector field in a neighborhood of P by

$$L = \frac{\partial\rho}{\partial z_2}(P)\frac{\partial}{\partial z_1} - \frac{\partial\rho}{\partial z_1}(P)\frac{\partial}{\partial z_2} .$$

Then L , $\bar{L}$ span (over $\mathbb{R}$) the <u>complex tangent space</u> to $\partial\Omega$ at points near P . Their span has no component in the <u>complex normal</u> direction

$$Z = \mathrm{Im}\left[\frac{\partial\rho}{\partial z_1}(P)\frac{\partial}{\partial z_1} + \frac{\partial\rho}{\partial z_2}(P)\frac{\partial}{\partial z_2}\right] .$$

However define

$$\mathcal{L}_0 = \mathrm{span}_{\mathbb{R}}\{L,\bar{L}\}$$
$$\mathcal{L}_1 = \mathrm{span}_{\mathbb{R}}\{\mathcal{L}_0,[\mathcal{L}_0,L],[\mathcal{L}_0,\bar{L}]\}$$

.

.

.

$$\mathcal{L}_j = \mathrm{span}_{\mathbb{R}}(\mathcal{L}_{j-1},[\mathcal{L}_{j-1},L],[\mathcal{L}_{j-1},\bar{L}]) .$$

We call P a point of finite type m if $\mathcal{L}_{m-1}$ contains no element with non-zero component in the direction Z while $\mathcal{L}_m$ does contain such an element.

Implicit in Kohn's paper [KO1] is the fact that the two definitions of finite type which we have given, one in terms of order of contact of non-singular varieties and the other in terms of commutators of vector fields, are equivalent. The main thrust of Kohn's paper [KO1] was to show that, in $\mathbb{C}^2$, finite type points P are precisely those near which a subelliptic estimate for the $\bar{\partial}$-Neumann problem of the form

$$|||u|||_\epsilon^2 \leqslant C[\|\bar{\partial}u\|^2 + \|\bar{\partial}^*u\|^2]$$

holds. Here u is a test function supported in a neighborhood of P, $|||u|||_\epsilon$ is a tangential Sobolev norm of order ϵ, and $\| \ \|$ is the 0-order Sobolev (or L^2) norm. Kohn estimated ϵ in terms of the type m and subsequent work in [GR] and [KR2] showed that this estimate is sharp.

Since 1972, there has been a great deal of work to determine the correct analogue of "finite type", to determine the sharp value for ϵ, and also to determine the right necessary and sufficient conditions for subelliptic estimates for the $\bar{\partial}$-Neumann problem, in dimensions exceeding two. Bloom and Graham [BG] formulated a definition of finite type (in any dimension) in terms of order of contact of complex hypersurfaces and proved this definition to be equivalent to one in terms of commutators of vector fields. While this notion of type was helpful to those thinking about peak points and holomorphic support functions (see [BL1],[BL2],[HS1]), it soon became clear that it was <u>not</u> the right condition for subelliptic estimates.

Kohn's important work in [KO2] gave a sufficient condition, in terms of ideals of forms, for the existence of subelliptic estimates for the $\bar{\partial}$-Neumann problem on forms in any dimension. He conjectured that his condition would also be necessary. At about the same time, D'Angelo and Catlin initiated a deep and protracted study of the program which Kohn initiated.

In a series of papers, D'Angelo developed from first principles an algebro-geometric theory of points of finite type. His semi-continuity result in [DAN] signalled that he had found the right theoretical framework. Meanwhile, Catlin built on D'Angelo's ideas and use his own deep insights into the construction of plurisubharmonic functions as a tool for attacking the problem of subelliptic estimates. From the work of D'Angelo and Catlin there evolved the following definition of finite type in $\mathbb{C}^n$:

<u>Definition</u>: Let $\Omega = \{z \in \mathbb{C}^n : \rho(z) < 0\}$ and $P \in \partial\Omega$. If $\gamma: \mathbb{C} \longrightarrow \mathbb{C}^n$ is holomorphic and $\gamma(0) = P$ then define

$$\tau(\gamma) = \frac{\nu(\rho\circ\gamma)}{\nu(\gamma)} ,$$

where $\nu(*)$ denotes the order of vanishing of $*$. Define the <u>type</u> $T(P)$ of P to be

$$T(P) \equiv \sup_{\gamma} \tau(\gamma) .$$

We say that P is of <u>finite</u> <u>type</u> if and only if $T(P) < \infty$.

Catlin [CAT2] has proved that if $\partial\Omega$ is pseudoconvex near P then a subelliptic estimate holds near P for the $\bar{\partial}$-Neumann problem if and only if P is of finite type. Diederich and Fornaess [DF3] have proved that a bounded pseudoconvex domain with real analytic boundary is of finite type. This means that, given a bounded domain Ω with real analytic boundary, there is a number $M > 0$ such that each point of $\partial\Omega$ has finite type not exceeding M . These results represent some of the most important progress made in several complex variables in the last fifteen years. They provide us with a large collection of domains on which the basic constructions of complex function theory can (at least in principle) be performed. Section 3 contains some particularly dramatic applications of these results.

It is my view that an important direction of future research ought to be the detailed study of harmonic analysis on domains of finite type. The development of Fatou theorems, Lusin area integrals, and admissible maximal functions in the special case of strongly pseudoconvex domains has already led to new understanding of singular integral operators, covering theorems, spaces of homogeneous type, and the other tools of harmonic analysis (see [ST1], [FOS1]). More recently, progress has been made on domains of finite type in $\mathbb{C}^2$ (see [NSW1],[KR4]). It is becoming increasingly clear that invariant metrics, such as the Bergman, Caratheodory, and Kobayashi metrics, will play a vital role in the final understanding of these issues ([KR3],[KR4]). A careful understanding of Catlin's work should lead to new developments on domains of finite type in $\mathbb{C}^n$, $n > 2$.

The papers of D'Angelo in this volume provide a valuable discussion of points of finite type, both from the point of view of commutators and the point of view of algebraic geometry. I hope that they will enable a new generation of researchers to consider (i) the behavior of invariant metrics near boundary points of finite type (see [CAT1] for results in $\mathbb{C}^2$), (ii) the theory of quadratic integrals, such as the Littlewood-Paley g function and the Lusin area integral, near ponts of finite type (see [ST1], [NSW1], [KR4]), (iii) boundary behavior of holomorphic functions near points of finite type (see [NSW1],[KR3], [KR4]), (iv) the theory of "real Hardy spaces" near points of finite type (see [FES], [FOS2]), (v) asymptotic expansions for canonical kernels (such as the Bergman, Szegö, and Neumann kernels) near poins of finite type (see [GRS], [NSW3]), and (vi) construction of non-canonical kernels (such as the Henkin-Ramirez kernel) near points of finite type (see [GRS], [FO2] for results in $\mathbb{C}^2$).

Section 3 The Fefferman Mapping Theorem and Related Results

In [FE1], C. Fefferman proved the following striking result:

Theorem: Let Ω_1 and Ω_2 be smoothly bounded, strictly pseudoconvex domains in $\mathbb{C}^n$. Let $\Phi: \Omega_1 \to \Omega_2$ be a biholomorphic map. Then Φ extends to a diffeomorphism of $\bar{\Omega}_1$ to $\bar{\Omega}_2$.

Fefferman's proof involves a detailed analysis of the asymptotic behavior of Bergman metric geodesics at the boundary and the details are too cumbersome to be presented here. More relevant for our purposes is a consideration of the impact of this theorem. Besides the solution of the Levi problem (and related topics such as the $\bar{\partial}$-Neumann problem), this was one of the very first theorems proved about a class of domains. It was certainly the first such result about holomorphic mappings. Prior to Fefferman, consideration of holomorphic mappings proceeded by explicit calculation of mappings of explicitly given domains described by polynomial inequalities. While these calculations were often quite difficult (see [HUA]) and involved powerful machinery (e.g. the Lie theory in the program of Cartan--see [HEL]), they represent what can now be safely called a classical chapter in complex analysis. Even though the details of Fefferman's proof are extremely difficult, it should be stressed that the statement of the theorem is readily accessible; moreover the very form of Fefferman's results already leads to new insights.

As an instance of this last remark, Klembeck [KL] used Fefferman's asymptotic formula for the Bergman kernel to calculate the asymptotic boundary behavior of the Bergman metric on a strongly pseudoconvex domain. One result is that the holomorphic sectional curvature of the metric approaches that of the ball. Coupled with a result of Lu Qi-Keng about complex manifolds of constant holomorphic sectional curvature, this yields a new proof of Bun Wong's theorem: the only strongly pseudoconvex domain with transitive automorphism group is the ball. Greene and Krantz ([GK1], [GK2], [GK3]), using Kelmbeck's work as inspiration, were able to do a more detailed analysis of Fefferman's asymptotic expansion and thus to learn how the group of biholomorphic maps of a domain depends on the boundary of that domain.

Probably the most profound consequence of Fefferman's theorem is that it vindicates a program of Poincaré to calculate (biholomorphic) differential boundary invariants for strongly pseudoconvex domains. A formal argument with power series ([FE1],[GK4]) shows that the invariants exist in principle, but one needs to know that biholomorphisms extend smoothly to the boundary before these "potential invariants" can be considered true invariants. Immediately following Fefferman's result, Chern and Moser [CM] completed Poincaré's program in principle in that they showed how the invariants can be calculated. Their results were to some extent anticipated by those of Tanaka in [TAN]. Burns, Shnider, and Wells ([BSW],[BS]) used the invariants to obtain important insights into the biholomorphic self-maps of domains. Perhaps the deepest work on the boundary invariants has been done by Fefferman himself in [FE2], [FE3]. In these papers Fefferman gave, among other things, an effective procedure for deciding whether two strongly pseudoconvex boundary points are biholomorphically equivalent. It should be stressed that Fefferman's results here, while very complete, apply only to biholomorphic equivalence <u>at a single point</u>. Much less is known about global obstructions to biholomorphic equivalence (however see [GK2], [GK3], [GK4],[BED1],[BDA]). For weakly pseudoconvex domains, even though there are many results about smoothness to the boundary of biholomorphic mappings, essentially nothing is known about biholomorphic differential invariants in the boundary. Even the case of finite type 3 in $\mathbb{C}^2$ has not been developed. Clearly there is much important work to be done in this area.

Fefferman's theorem has also inspired many people to consider both extending and simplifying the biholomorphic mapping theorem. A very detailed survey of this work was given by Bedford in [BED2], and I shall only make a few remarks about it here.

The explicit nature of Fefferman's asymptotic expansion required the creation of a delicate calculus of singular integrals (Boutet de

Monvel and Sjöstrand [BMS] succeeded in making this part of the proof more natural by fitting it into the context of Fourier integral operators). Clearly one goal in simplifying Fefferman's work would be to bypass the asymptotic expansion (which has great value in other contexts) and to find a proof of the biholomorphic mapping theorem which hinges on softer properties of the Bergman kernel. On the other hand, Fefferman's analysis of Bergman metric geodesics has some very difficult points in it and there is interest in finding a way around that part of the proof. Several people rose to this challenge and have provided us with new insights.

S. Webster, in his paper [WE], discovered that the Bergman kernel could be used to create a special coordinate system in which biholomorphic mappings become trivial. While his ideas are in a sense a rediscovery of Bergman representative coordinates, they are in another way quite profound for their apparent connection with the boundary behavior of biholomorpic mappings (Bergman's ideas were purely formal and really only apply on the interior). Nirenberg, Webster, and Yang [NWY] gave a method, which in my opinion has been neither fully explored nor developed, of reducing smoothness to the boundary of biholomorphic mappings to the problem of showing that such mappings are $C^{1+\epsilon}$ to the boundary. There is a marriage of several nice ideas here. On the one hand Yang's theory of horocycles gives a certain amount of boundary smoothness from below. On the other hand, Nirenberg's reflection argument gives the bootstrap argument that extracts C^∞ regularity from $C^{1+\epsilon}$ regularity. Together the three authors made a complete argument from these pieces. I would not be surprised if this sort of argument worked in greater generality.

Surely the most far-reaching program to extend and simplify Fefferman's theorem has been that initiated by Bell/Ligocka [BELL] and Bell [BEL]. The single most important contribution to come from this program is the discovery of the central role of the mapping properties of the Bergman projection in the study of biholomorphic maps. Bell's Condition R (for a domain Ω) is the following: If $K(z,\zeta)$ is the Bergman kernel for Ω then the mapping

$$P_\Omega: f \longmapsto \int K(z,\zeta)\, f(\zeta)\, dV(\zeta)$$

should be bounded from $C^\infty(\bar{\Omega})$ to $C^\infty(\bar{\Omega})$. One of Bell's theorems says that if Ω_1 and Ω_2 are smooth pseudoconvex domains and one of them has Condition R then any biholomorphic mapping of Ω_1 to Ω_2 extends to a diffeomorphism of the closures.

Bell and Bell/Ligocka have succeeded in eliminating both the asymptotic expansion for the Bergman kernel <u>and</u> the differential geometry (study of geodesics) from the proof of the mapping theorem. The standard method (though by no means the only method) for verifying Condition R for a domain Ω is by way of subelliptic estimates for the $\bar{\partial}$-Neumann problem. Kerzman [KER] first noticed the connection between the kernel and the $\bar{\partial}$-Neumann problem; he also introduced a beautiful trick for reading off information about the Bergman kernel from its mapping properties and vice versa. Bell was able to exploit these by making use of the following projection formula:

<u>Proposition ([BEL])</u>: Let $\varphi: \Omega_1 \longrightarrow \Omega_2$ be biholomorphic, $u = \det \mathrm{Jac}_{\mathbb{C}}\varphi$, and $g \in L^2(\Omega_2)$. Then

$$P_1(u\cdot(g\circ\varphi)) = u\cdot((P_2(g))\circ\varphi) .$$

This projection formula is the key to getting from Condition R to the mapping theorem. Bell's ideas have produced a new conceptual framework in which to consider biholomorphic mappings. It added impetus to Kohn's program of learning which domains have subelliptic estimates for the $\bar{\partial}$-Neumann problem. Thanks to Catlin's work, we now know that a biholomorphic mapping of two domains, one of which is pseudoconvex and of finite type, extends to a diffeomorphism of the closures.

It has been conjectured that the mapping theorem is true for arbitrary smoothly bounded domains Ω_1 and Ω_2, just as it is in one complex dimension. In optimistic moments, it has even been conjectured that all smoothly bounded domains satisfy Condition R. Barrett [BAR1] produced a counterexample to this last conjecture, but his counterexample is <u>not</u> pseudoconvex. There are also examples, in [WE] and [FR], of biholomorphisms of pseudoconvex domains which do not continue C^1, or even C^0, to the boundary; however in these counterexamples the domains do <u>not</u> have C^2 boundary. We still hope that when the boundary is at least C^2 then biholomrphic mappings will have a predictable amount of smoothness (after all, many geometric constructions only work when the boundary is C^2). On the other hand, this may turn out to be a situation like the Nebenhülle problem: we may be trying to prove the wrong thing.

The examples of [WE] and [FR] have certainly sharpened our focus, but the possibility still remains that a biholomorphic mapping of smoothly bounded domains will extend diffeomorphically to the closures. If this conjecture is correct, then it will clearly <u>not</u> depend on subelliptic estimates. Indeed subelliptic estimates, being a delicate local property of the partial differential operator $\bar{\partial}$, are probably too powerful a tool for the problem at hand. What may be more suitable forattacking this conjecture is softer <u>global</u> arguments. Bolstering this point of view is the recent paper of Barrett [BAR2] in which he proves that there is no local obstruction to Condition R: any obstruction, if it exists, is global.

If we restrict attention to smoothly bounded pseudoconvex domains Ω, then the paper [KO3] of Kohn comes to mind. There Kohn used soft global arguments to show that if f is a $\bar{\partial}$-closed form with smooth coefficients on the closure of a smoothly bounded pseudoconvex domain Ω then there is a u with smooth coefficients on $\bar{\Omega}$ such that $\bar{\partial}u = f$. The solution u <u>is</u> the Neumann solution to the $\bar{\partial}$ problem, but <u>not in the usual metric</u>.

Thus while Kohn's result is a version of Condition R, it is not Condition R for the usual Bergman projection, but rather the projection in a different metric. Unfortunately, there is no analogue for Bell's projection formula in the metric that Kohn uses, and the program breaks down. Nevertheless, this scheme suggests the sort of global argument which may someday work in general.

Another encouraging development is the recent work of Baouendi, Jacobowitz, and Treves [BJT] which shows that a biholomorphic mapping of any two pseudoconvex domains with real analytic boundaries will continue analytically past the boundaries. Prior to their work, very little was known beyond the strongly pseudoconvex case (however see the work of Moser and Webster in [MOW]). It is very encouraging to encounter theorems, like the ones in [BJT], which work "all the time". We should take special note of the fact that the techniques of that paper, especially the use of the Fourier-Bros-Iagnolltzer (or FBI) transform, were completely unknown to people who had been working on mapping problems. It may be that other methods of partial differential equations, besides the by now well-known ones connected with the $\bar{\partial}$-Neumann problem and complex Monge-Ampère equation, are relevant to mapping problems. As far as I know, nobody has investigated this possibility.

Finally let me mention that the techniques that have been described in this section have also been brought to bear on the study of <u>proper</u> holomorphic mappings. We have learned that, for many purposes, proper maps are as tractable as biholomorphic maps. And the study of this broader problem has shed light on the original biholomorphic mapping problem. For example, a recent development is the discovery of variants of the Schwarz lemma at the boundary for biholomorphic and proper mappings (see Bell's article in this volume, as well as [KR5], [GK5]). These have in turn revealed interesting connections with interpolation problems for H^∞ functions, and should prove useful in a variety of contexts.

The study of proper mappings has given new vitality to this branch of the field because, for instance, it has established connections with division problems (see [DF4],[BEC]) and has led to the definition and study of holomorphic <u>correspondences</u> (see [BB]). Again, Bedford's survey [BED2] is a good source of information on this material.

<u>Section 4</u> Concluding Remarks

It is easy to look back on the progress of the last fifteen years and, fortified by a feeling of much excellent work completed, wonder what there is left for the next fifteen years. It has been suggested recently by many people, including myself, that we are running out of things to do. After listening to the lectures at this conference, and after writing this article, I no longer feel that way.

What is true, I think, is that our perspective now is very different than it was in 1970. In 1970 we had no detailed information about weakly pseudoconvex domains. Strongly pseudoconvex domains were only barely understood. It is unlikely that in the next several years there will be results which are as earthshaking as were the worm domain or the bounded plurisubharmonic exhaustion function, primarily because we are now more experienced and realize what we can and cannot expect. If someone were to solve the corona problem for the ball and the biholomorphic mapping problem for arbitrary domains in 1988 then, great achievements that these would be, they would not have the impact that Fefferman's mapping theorem had in 1974. As we begin to understand a subject better, we become more jaded.

I hope that this essay suggests that several complex variables is not running out of steam, but rather that it is richer than we ever suspected. Now that we have reached a first plateau of maturity, we can look back and see a continuing historical dynamic:

> Elementary consideration of strong pseudoconvexity and analytic continuation led to the Lewy unsolvable partial differential operator. What new insights will a better understanding of weak pseudoconvexity reveal?
>
> A neat observation of Kerzman gave a connection between the Bergman kernel and the $\bar{\partial}$-Neumann problem; this in turn led to Condition R and eventually to connections among the mapping problem, subellipticity for the $\bar{\partial}$-Neumann problem, and points of finite type. What new insights will a solution of the mapping problem bring?

We now know a hand full of domains on which the $\bar{\partial}$-Neumann problem satisfies uniform estimates; Sibony [SI3] has exhibited a smoothly bounded pseudoconvex domains on which there are no uniform estimates. The determination of necessary and sufficient conditions for uniform estimates for the $\bar{\partial}$ problem is bound to give insight into the nature of pseudoconvexity.

I have already alluded to the role of invariant metrics in harmonic analysis questions. It is also clear, both from Fefferman's work in [FE1] and from work of Henkin [HEN2], Range [RAN1], Diederich/Fornaess [DF4] and others, that invariant metrics explain much of what is going on in mapping problems. It is a deep and difficult problem to calculate invariant metrics on general types of domains, and this certainly should be one of the new frontiers of research.

I think that the many new connections between several complex variables and the subjects of differential geometry, partial differential equations, harmonic analysis, and Lie group theory need to be further explored. As the study of the $\bar{\partial}$-Neumann problem, points of finite type, and the mapping problem have shown us, all these disciplines can lead to important new discoveries in several complex variables. Cirka [CI] once described (a part of) several complex variables as "an entire field of white nothingness with widely spaced isolated results". His description may still be accurate, but there are now a lot more tools at hand, and a lot more knowledge and experience, to bring to bear on the subject.

Probably the most exciting conference that I ever attended was the 1975 Williamstown conference on several complex variables. There had not been a conference in the subject for some time, and there were an enormous number of new developments: integral formulas for strongly pseudoconvex domains, Fefferman's mapping theorem, the Chern-Moser invariants, the work of Henkin and Skoda on zero sets of Nevanlinna functions, sharp estimates for the $\bar{\partial}$ problem, the worm domain, bounded plurisubharmonic exhaustion functions, the Bloom/Graham work on points of finite type, and so on. I think that those who attended were profoundly influenced by what went on. This was not merely a bunch of the boys giving the same talks that we heard last year. For many of us, the talks were a revelation.

However, out of everything that I heard during those three weeks, the most memorable is a single sentence spoken by Stefan Bergman. In the middle of one of the twenty minute talks, Bergman stood up and said "I think that you people studying biholomorphic mappings ought to look at representative coordinates". We didn't pay much attention at the time--in fact, almost nobody knew what he was talking about. But he was right, as we learned five years later. Bergman representative coordinates were rediscovered by several different people, and they were used (after suitable modification) to great effect in the programs of Webster, Bell and Ligocka, and Greene and Krantz. I do not think that their full utility has yet been realized.

I wonder how many other powerful old techniques we have forgotten? I don't know anyone who has studied the books of Behnke and Thullen [BTU], Bochner and Martin [BOM], Cartan [CAR], B. A. Fuks [FUK1],[FUK2], (the latter contains the only textbook discussion of representative coordinates), Forsyth [FOR], Malgrange [MAL], or Vladimirov [VLA]. As the recent publication of the collected works of Oka [OKA] reveals, it is very painful to read the writings of the old workers in several complex variables. The language has completely changed (to all appearances, for the better). But there must be some great ideas buried in that arcane language. The 1906 paper [HAR] of Hartogs is 88 pages long--surely it contains a lot more than the Hartogs extension phenomenon. The old papers of Bergman (see [BERG]) contain geometric descriptions of domains in $\mathbb{C}^2$ and $\mathbb{C}^3$ which defy translation into modern language. I wish that I could understand them.

In some sense, the development of powerhouse techniques like sheaf theory, the $\bar{\partial}$ problem, and Kähler geometry, much of which took place between 1945 and 1965, managed to cut us off from the roots of several complex variables. Very little is remembered about the techniques of the first part of the century, which happens to be the time when the foundations of the subject were laid. I think that one of the good features of the work since 1970 is that many of the best papers rely little on powerhouse methods; instead, they involve calculation and experimentation. When I read the papers on the worm domain or the Kohn-Nirenberg example or points of finite type, I am reminded of the old work of E. E. Levi in which he grappled, at the most basic level, with what pseudoconvexity ought to be. It could be argued that, in the last fifteen years, more has been accomplished with "elementary" methods than with sheaf theory.

I certainly do not advocate abandoning the beautiful machines that have been developed. What I do advocate is a reinvestigation of the basic ideas of the subject--pseudoconvexity, domains of holomorphy, etc.--and especially a re-reading of the old literature. Even if reading the work of Hartogs and Oka doesn't help us solve any current problems, it is bound to suggest a lot of new ones--new, at least, to us.

The author gratefully acknowledges partial support from the National Science Foundation.

REFERENCES

[BJT] S. Baouendi, H. Jacobowitz, and F. Treves, On the analyticity of CR mappings, preprint.

[BAR1] D. Barrett, Irregularity of the Bertgman projection on a smooth, bounded domain in $\mathbb{C}^2$, Ann. Math. 119(1984), 431-436.

[BAR2] D. Barrett, Regularity of the Bergman projeciton and local geometry of domains, Duke Jour. Math. 53(1986), 333-343.

[BED1] E. Bedford, Invariant forms on complex manifolds with applications to holomorphic mappings, Math. Annalen 265(1983), 377-396.

[BED2] E. Bedford, Proper holomorphic mappings, Bull. Am. Math. Soc. 10(1984), 157-175.

[BB] E. Bedford and S. Bell, Boundary continuity of proper holomorphic correspondences, Seminaire Dolbeault - Lelong - Skoda, 1983.

[BDA] E. Bedford and J. Dadok, Bounded domains with prescribed automorphism groups, preprint.

[BEF] E. Bedford and J. E. Fornaess, A construction of peak functions on weakly pseudoconvex domains, Ann. Math. 107(1978), 555-568.

[BTU] H. Behnke and P. Thullen, Theorie der Funktionen Mehrerer Komplexer Veränderlichen, Second Edition, Springer, Berlin, 1970.

[BEL] S. Bell, Biholomorphic mappings and the $\bar\partial$ problem, Ann. Math. 114(1981), 103-112.

[BEC] S. Bell and D. Catlin, Proper holomorphic mappings extend smoothly to the boundary, Bull. Am. Math. Soc. 7(1982), 269-272.

[BELL] S. Bell and E. Ligocka, A simplification and extension of Fefferman's theorem on biholomorphic mappings, Invent. Math. 57(1980), 283-289.

[BERG] S. Bergman, The kernel function and conformal mapping, Am. Math. Soc., Providence, 1970.

[BER] L. Bers, Introduction to Several Complex Variables, New York University Press, New York, 1964.

[BL1] T. Bloom, C^∞ peak functionns for pseudoconvex domains of strict type, Duke Math. J. 45(1978), 133-147.

[BL2] T. Bloom, On the contact between complex manifolds and real hypersurfaces in $\mathbb{C}^3$, Trans. Am. Math. Soc. 263(1981), 515-529.

[BG] T. Bloom and I. Graham, A geometric characterization of points of finite type m on real submanifolds of $\mathbb{C}^n$, J. Diff. Geom. 12(1977), 171-182.

[BOM] S. Bochner and W. Martin, Several Complex Variables, Princeton University Press, Princeton, 1948.

[BMS] L. Boutet de Monvel and J. Sjöstrand, Sur la singularité des noyaux de Bergman et Szegö, Soc. Mat. de France Asterisque 34-35(1976), 123-164.

[BS] D. Burns and S. Shnider, Geometry of hypersurfaces and mapping theorems in $\mathbb{C}^n$, Comment. Math. Helv. 54(1979), 199-217.

[BSW] D. Burns, S. Shnider, and R. Wells, On deformations of strongly pseudoconvex domains, Invent. Math. 46(1978), 237-253.

[CAR] H. Cartan, Théorie élémentaire des fonctions analytiques d'une ou plusieurs variables complexes, Hermann, Paris, 1964.

[CAT1] D. Catlin, Invariant metrics on pseudoconvex domains, in Several Complex Variables: Proceedings of the 1981 Hangzhou conference, ed. J. J. Kohn, Q-K Lu, R. Remmert, Y. T. Siu, Birkhäuser, Boston, 1984.

[CAT2] D. Catlin, Subelliptic estimates for the $\bar{\partial}$ - Neumann problem on pseudoconvex domains, preprint.

[CM] S. S. Chern and J. Moser, Real hypersurfaces in complex manifolds, Acta Math. 133(1974), 219-271.

[CI] E. Cirka, The theorems of Lindelöf and Fatou in $\mathbb{C}^n$, Mat. Sb. 92(134)(1973), 622-644; Math. USSR Sb. 21(1973), 619-639.

[DAN] J. D'Angelo, Real hypersurfaces, orders of contact and applications, Ann. Math. 115(1982), 615-638.

[DF1] K. Diederich and J. E. Fornaess, Pseudoconvex domains: an example with nontrivial Nebenhülle, Math. Annalen 225(1977), 272-292.

[DF2] K. Diederich and J. E. Fornaess, Pseudoconvex domains: bounded plurisubharmonic exhaustion functions, Invent. Math. 39(1977), 129-141.

[DF3] K. Diederich and J. E. Fornaess, Pseudoconvex domains with real analytic boundaries, Ann. Math. 107(1978), 371-384.

[DF4] K. Diederich and J. E. Fornaess, Biholomorphic mappings between certain real analytic domains in $\mathbb{C}^n$, Math. Annalen 245(1979), 255-272.

[DF5] K. Diederich and J. E. Fornaess, Smooth extendability of proper holomorphic mappings, Bull. Am. Math. Soc. 7(1982), 264-268.

[FE1] C. Fefferman, The Bergman kernel and biholomorphic mappings of pseudoconvex domains, Invent. Math. 26(1974), 1-65.

[FE2] C. Fefferman, Monge Ampère equations, the Bergman kernel, and the geometry of pseudoconvex domains, Ann. Math. 103(1976), 395-416.

[FE3] C. Fefferman, Parabolic invariant theory in complex analysis, Adv. Math 31(1979), 131-262.

[FES] C. Fefferman and E. M. Stein, H^p spaces of several variables, Acta Math. 129(1972), 137-193.

[FOK] G. B. Folland and J. J. Kohn, The Neumann Problem for the Cauchy-Riemann Complex, Princeton University Press, Princeton, 1972.

[FOS1] G. B. Folland and E. M. Stein, Estimates for the $\bar{\partial}_b$ complex and analysis of the Heisenberg group, Comm. Pure and Appl. Math. 27(1974), 429-522.

[FOS2] G. B. Folland and E. M. Stein, Hardy spaces on Homogeneous Groups, Math. Notes 28, Princeton University Press, Princeton, 1982.

[FO1] J. E. Fornaess, Peak points on weakly pseudoconvex domains, Math. Annalen 227(1977), 173-175.

[FO2] J. E. Fornaess, Sup norm estimates for $\bar{\partial}$ in $\mathbb{C}^2$, preprint.

[FOR] A. R. Forsyth, Lectures Introductory to the Theory of Two Complex Variables, Cambridge University Press, Cambridge, 1914.

[FR] B. Fridman, One example of the boundary behavior of biholomorphic transformations, preprint.

[FUK1] B. A. Fuks, Introduction to the Theory of Analytic Functions of Several Complex Variables, Translations of Mathematical Monographs, American Math. Society, Providence, 1963.

[FUK2] B. A. Fuks, Special Chapters in the Theory of Analytic Functions of Several Complex Variables, Translations of Mathematical Monographs, American Math. Society, Providence, 1965.

[GK1] R. E. Greene and S. G. Krantz, Stability of the Bergman kernel and curvature properties of bounded domains, in Recent Developments in Several Complex Variables. Annals of Math. Studies 100, Princeton University Press, Princeton, 1981.

[GK2] R. E. Greene and S. G. Krantz, Deformations of complex structures, estimates for the $\bar{\partial}$ equation, and stability of the Bergman kernel, Adv. in Math. 43(1982), 1-86.

[GK3] R. E. Greene and S. G. Krantz, The automorphism groups of strongly pseudoconvex domains, Math. Annalen 261(1982), 425-446.

[GK4] R. E. Greene and S. G. Krantz, Biholomorphic self-maps of domains, Proceedings of a Special Year in Complex Analysis at the University of Maryland, Springer Lecture Notes, Springer,

1987.

[GK5] R. E. Greene and S. G. Krantz, A new invariant metric in complex analysis and some applications, to appear.

[GR] P. Greiner, Subelliptic estimates for the $\bar{\partial}$-Neumann problem in $\mathbb{C}^2$, J. Diff. Geom. 9(1974), 239-250.

[GRS] P. Greiner and E. M. Stein, On the solvability of some differential operators of type $\square_b$, Proceedings of International Conferences, Cortona, Italy, Scuola Normale Superiore Pisa, 1978.

[HS1] M. Hakim and N. Sibony, Quelques conditions pour l'existence de fonctions pics dans les domains pseudoconvexes, Duke Math. Jour. 44(1977), 399-406.

[HS2] M. Hakim and N. Sibony, Frontière de Šilov et spectre de A($\bar{D}$) pour des domaines faiblement pseudoconvex, C. R. Acad. Sci. Paris, Sér. A-B 281(1975), A959-A962.

[HAR] F. Hartogs, Zur Theorie der analytischen Functionen mehrener unabhänghiger Veränderlichen insbesondere über die Darstellung derselben durch Reihen, welche nach Potenzen einer Veränderlichen fortschreiten, Math. Annalen 62(1906), 1-88.

[HEL] S. Helgason, Differential Geometry and Symmetric Spaces, Academic Press, New York, 1962.

[HEN1] G. M. Henkin, Integral representations of functions holomorphic in strictly pseudoconvex domains and some applications, Mat. Sb. 78(120)(1969), 611-632; Math. USSR Sb. 7(1969), 597-616.

[HEN2] G. M. Henkin, An analytic polyhedron is not holomorphically equivalent to a strictly pseudoconvex domain (Russian), Dokl. Acad. Nauk. SSSR 210(1973), 1026-1029.

[HOR] L. Hörmander, Introduction to Complex Analysis in Several Variables, North Holland, Amsterdam, 1973.

[HUA] L. K. Hua, Harmonic Analysis of Functions of Several Complex Variables in the Classical Domains, American Mathematical Society, Providence, 1963.

[KER] N. Kerzman, The Bergman kernel function. Differentiability at the boundary, Math. Annalen 195(1972), 149-158.

[KL] P. Klembeck, Kähler metrics of negative curvature, the Bergman metric near the boundary and the Kobayashi metric on smooth bounded strictly pseudoconvex domains sets, Indiana Univ. Math. J. 27(1978), 275-282.

[KO1] J. J. Kohn, Boundary behavior of $\bar{\partial}$ on weakly pseudoconvex manifolds of dimension two, J. Diff. Geom. 6(1972), 523-542.

[KO2] J. J. Kohn, Sufficient conditions for subellipticity of weakly pseudoconvex domains, Proc. Nat. Acad. Sci. (USA) 74(1977), 2214-2216.

[KO3] J. J. Kohn, Global regularity for $\bar{\partial}$ on weakly pseudoconvex manifolds, Trans. Am. Math. Soc. 181(1973), 273-292.

[KON] J. J. Kohn and L. Nirenberg, A pseudo-convex domain not admitting a holomorphic support function, Math. Annalen 201(1973), 265-268.

[KR1] S. Krantz, Function Theory of Several Complex Variables, John Wiley and Sons, New York, 1982.

[KR2] S. Krantz, Characterizations of various domains of holomorphy via $\bar{\partial}$ estimates and applications to a problem of Kohn, Ill. J. Math. 23(1979), 267-286.

[KR3] S. Krantz, Fatou theorems on domains in $\mathbb{C}^n$, Bull. Am. Math. Soc., to appear.

[KR4] S. Krantz, Invariant metrics and harmonic analysis on domains in $\mathbb{C}^n$, to appear.

[KR5] S. Krantz, A compactness principle complex analysis, Division de Matematicas U. A. M. Seminarios, to appear.

[MAL] B. Malgrange, Lectures on the Theory of Functions of Several Complex Variables, Tata Institute of Fundamental Research, Bombay, 1958.

[MOW] J. Moser and S. Webster, Normal forms for real surfaces in $\mathbb{C}^2$ near complex tangents and hyperbolic surface transformations, Acta Math. 150(1983), 255-298.

[NSW1] A. Nagel, E. M. Stein, and S. Wainger, Boundary behavior of functions holomorphic in domains of finite type, Proc. Nat. Acad. Sci. (USA) 78 (1981), 6596-6599.

[NSW2] A. Nagel, E. M. Stein, and S. Wainger, Balls and metrics defined by vector fields I: Basic properties, Acta Math. 155(1985), 103-148.

[NSW3] A. Nagel, oral communication.

[NWY] L. Nirenberg, S. Webster, and P. Yang, Local boundary behavior of holomorphic mappings, Comm. Pure Appl. Math. 33(1980), 305-338.

[OKA] K. Oka, Collected Papers, Springer, Berlin, 1984.

[RAM] E. Ramirez, Divisions problem in der Komplexen analysis mit einer Anwendung auf Rand integral darstellung, Math. Ann. 184(1970), 172-187.

[RAN1] R. M. Range, The Caratheodory metric and holomorphic maps on a class of weakly pseudoconvex domains, Pac. Jour. Math. 78(1978), 173-189.

[RAN2] R. M. Range, Holomorphic Functions and Integral Representations in Several Complex Variables, Springer Verlag, New York, 1986.

[SI1] N. Sibony, Sur le plongement des domaines faiblement pseudoconvexes dans des domaines convex, Math. Ann. 273(1986), 209-214.

[SI2] N. Sibony, Un exemple de domaine pseudonconvexe regulier ou l'equation $\bar{\partial}u = f$ n'admet pas de solution bornee pour f bornee, Invent. Math. 62(1980), 235-242.

[ST1] E. M. Stein, Boundary Behavior of Holomorphic Functions of Several Complex Variables, Princeton University Press, Princeton, 1972.

[ST2] E. M. Stein, Singular Integrals and Differentiability Properties of Functions, Princeton University Press, Princeton, 1970.

[TA] N. Tanaka, On generalized graded Lie algebras and geometric structures I, J. Math. Soc. Japan 19(1967), 215-254.

[VLA] V. Vladimirov, Methods of the Theory of Functions of Many Complex Variables, MIT Press, Cambridge, 1964.

[WE] S. Webster, Biholomorphic mappings and the Bergman kernel off the diagonal, Invent. Math. 51(1979), 155-169.

BOUNDARY SINGULARITIES OF BIHOLOMORPHIC MAPS

David E. Barrett
Dept. of Mathematics
Princeton University
Princeton, NJ 08544

Let D_1 and D_2 be relatively compact domains with smooth boundaries contained in the complex manifolds M_1 and M_2, respectively, and suppose that there is a biholomorphic map F from D_1 to D_2. Let Γ denote the graph of F and let $\overline{\Gamma}$ denote the closure of Γ in $\overline{D}_1 \times \overline{D}_2$. The <u>singular support</u> of F (denoted sing supp F) will be defined as the set of all points (p,q) in $\overline{\Gamma}\setminus\Gamma$ such that F cannot be extended to a diffeomorphism from a neighborhood of p in $\overline{D}_1$ to a neighborhood of q in $\overline{D}_2$. Thus sing supp F is a closed subset of $bD_1 \times bD_2$. If M_1 and M_2 are Stein and D_1 is a strictly pseudoconvex domain or, more generally, a pseudoconvex domain of finite type then it is known that sing supp F must be empty ([Be], [BBC],[C]). The following question is still open.

<u>Question 1.</u> *Must sing supp F = ∅ whenever M_1 and M_2 are Stein?*

It would be nice to resolve this question in particular for the important special case where $M_1 = M_2 = \mathbb{C}^n$ and D_1 and D_2 are weakly pseudoconvex.

We shall see by examples below that sing supp F need not be empty when M_1 and M_2 are not Stein. Nevertheless sing supp F must satisfy certain restrictions, as we see by the following result.

<u>Theorem ([Ba1]).</u> *If D_1 and D_2 are pseudoconvex, or if both domains have real analytic boundary, then sing supp F has no isolated points.*

(Observe that if (p,q) is an isolated point of sing supp F then it is immediate that F extends to a <u>homeomorphism</u> from a neighborhood of p in $\overline{D}_1$ onto a neighborhood of q in $\overline{D}_2$.)

The technique used in [Ba1] to prove the above theorem can be extended somewhat to produce other examples of sets which are too small to contain components of sing supp F. Note also that by shrinking the domains D_1 and D_2 we may see that an affirmative answer to Question 1 would imply in particular that no component of sing supp F is contained in a product $U_1 \times U_2$, where U_1 and U_2 are Stein open subsets of M_1 and M_2, respectively.

Given, then, that boundary singularities of biholomorphic maps do occur and that they must satisfy some restrictions it seems desirable to attempt to search for the rudiments of a structure theory of such singularities. A necessary step in this direction is to study some examples. In Section 1 below we exhibit examples of biholomorphic maps with singular boundary behavior; in particular we generalize the examples of [Ba2]. In Section 2 we pose some questions raised by these examples.

§1. Examples.

The simplest examples of biholomorphic maps with singular boundary behavior appear to be those which come from domains which admit a complex family of automorphisms. Consider, for instance, the manifold $M=\mathbb{C}\times\mathbb{P}^1$ and the domain $D=U\times\mathbb{P}^1$ where U is the unit disc. Any function f on D induces an automorphism F of D by the formula $F(z,w)=(z,w+f(z))$, where w is the standard inhomogeneous coordinate on $\mathbb{P}^1$. If f has singular boundary behaviour then so does F.

More generally, let M be a complex manifold which admits a $\mathbb{C}$-action such that the map $\mathbb{C}\times M \to M$, $(t,z) \mapsto t\cdot z$ is jointly holomorphic in t and z. Suppose that M contains a $\mathbb{C}$-invariant relatively compact domain D with smooth boundary and suppose further that D admits a $\mathbb{C}$-invariant holomorphic function f with singular boundary behavior. Let F be the automorphism of D given by $F(z) = F(f(z)\cdot z)$ and let V be the variety $\{ z\in M : t\cdot z=z \text{ for all } t \}$. Then by linearizing the $\mathbb{C}$-action locally at points of $M\times V$, where V is the variety $\{z\in M : t\cdot z=z \text{ for all } t\}$, it is easy to see that sing supp F contains at least the set

$$\{ (z,t\cdot z) : z \in (\text{sing supp } f)\setminus V,\ t \in \text{cluster set of } f \text{ at } z \}.$$

We could of course replace $\mathbb{C}$ by a more general complex Lie group in this set-up.

To construct examples which do not fall within the framework above we begin with the following lemma.

Lemma. *Let k be a positive integer and let Φ denote the self-map of the complex plane given by $\Phi(z)=z|z|^{2k}$. (Φ is a homeomorphism but neither a biholomorphism nor a diffeomorphism.) Let Ω be a bounded* *planar domain with smooth boundary. Then the domain* $\Phi^{-1}\Omega$ *also has smooth boundary. Furthermore if* Ω *has order of contact* s *with its tangent line at* 0 *then the corresponding order of contact for* $\Phi^{-1}b\Omega$ *is* $s+2k(s-1)$.

Proof. We may assume that $0 \in b\Omega$ and it will suffice to restrict our attention to a neighborhood of 0. After a rotation we may assume that $b\Omega$ is given near 0 by the equation $y+g(x)=0$, where g is smooth and vanishes to order s at 0. Then $b\Phi^{-1}\Omega$ is given near 0 by the equation $y+|z|^{-2k}g(|z|^{2k}x)=0$. Using Taylor's theorem it is easy to check that the function $y+|z|^{-2k}g(|z|^{2k}x)$ is smooth and has non-vanishing gradient at 0, and furthermore that the first non-vanishing term of its Taylor series which does not involve y is of order $s+2k(s-1)$. □

In order to use the above lemma to construct biholomorphic maps we first let M_2 be the quotient of $\mathbb{C}\times\mathbb{C}^*$ by the properly discontinuous fixed point-free group of automorphisms generated by

$$(z,w) \mapsto (\alpha z, \alpha w),$$

where α is a real number greater than 1; we continue to write points of M_2 in the form (z,w). Then the map

$$\Psi: M_2 \to \mathbb{C},\ (z,w) \mapsto |w|^{-1}z$$

is a well-defined submersion so that the relatively compact domain $D_2 = \Psi^{-1}\Omega$ has smooth boundary, where Ω is any domain satisfying the hypotheses of the Lemma. (We shall assume that $0 \in b\Omega$.)

Now pick a positive integer k and let M_1 be the manifold obtained from M_2 by dividing by the additional automorphisms

$$(z,w) \mapsto (z, \gamma^j w),\ j=0, \ldots, 2k,$$

where γ is a primitive $(2k+1)^{st}$ root of unity. M_1 is biholomorphic to M_2 via the map

$$F: M_1 \to M_2,\ (z,w) \mapsto (z, z^{-2k}w^{2k+1}).$$

Let $\tilde{\Psi}$ be the submersion $M_1 \to \mathbb{C}$, $(z,w) \mapsto |w|^{-1}z$. Then the relatively compact domain

$$D_1 = F^{-1}D_2 = F^{-1}\Psi^{-1}\Omega = \tilde{\Psi}^{-1}\Phi^{-1}\Omega$$

has smooth boundary by the Lemma.

Thus $F: D_1 \to D_2$ is a biholomorphic map of the sort we wish to consider, and by looking at the images of rays of the form $(\rho e^{i\theta}, w)$ with θ and w fixed and ρ decreasing to zero one can check that sing supp $F = \{ ((z_1,w_1),(z_2,w_2)) : z_1=z_2=0 \}$; i.e., sing supp F is a product of two elliptic curves.

Borrowing notation from the proof of the Lemma we note that D_2 is defined near the points where $z=0$ by the equation $r(z,w)=0$, where $r(z,w)=y+|w|g(|w|^{-1}x)$ with $z=x+iy$. Recalling that the Levi form of bD_2 is given up to normalizations by

$$-\det \begin{pmatrix} r & r_z & r_w \\ r_{\bar{z}} & r_{z\bar{z}} & r_{w\bar{z}} \\ r_{\bar{w}} & r_{z\bar{w}} & r_{w\bar{w}} \end{pmatrix}$$

it is easy to check that the order of vanishing of the Levi form of bD_2 at points where $z=0$ is the number s from the Lemma. Similarly, the corresponding order of vanishing for D_1 is $s+2k(s-1)$.

The geometry of D_1 and D_2 at points where $z\neq 0$ is best understood by observing that after choosing a branch of $\log z$ the map $G: (z,w) \mapsto (w^{-1}z, z^{-2\pi i/(\log \alpha)})$ is a well-defined biholomorphic map from D_2 onto a component D_3 of the region $\{ (z,w) : \exp(\log|z| + i(2\pi)^{-1}\log \alpha \cdot \log|w|) \in \Omega \}$ in $\mathbb{C}^2$; D_3 is of course just the Reinhardt domain whose logarithmic hull in $\mathbb{R}^2$ is the planar domain obtained by applying a branch of the complex logarithm function to the domain Ω and then stretching. The map G is well-behaved away from $z=0$.

The examples in [Ba2] are obtained (after a change of coordinates) by taking Ω to be a disc.

§2. Questions.

Let F be as in the introduction. On the basis of the rather limited supply of known examples with sing supp $F \neq \emptyset$ it is tempting to ask if the singularities of such a map must propagate along varieties; more precisely, we pose the following question.

Question 2. *Must sing supp F be locally the union of complex analytic varieties of positive dimension?*

Note that the maximum principle for strictly plurisubharmonic functions shows that an affirmative answer to Question 2 would imply an affirmative answer to Question 1.

At this writing the author is unsure if the answer to Question 2 is affirmative for all examples falling within the framework of the opening paragraphs of Section 1 above.

Note also that automorphisms play a role in both types of example considered in Section 1 above. (In the latter case we have an action of the two-torus $\mathbb{T}^2$ on D_2 given by $(\theta,\tau)\to(z,w)=(\alpha^{\tau/2\pi}z,\alpha^{\tau/2\pi}e^{i\theta}w)$ and a corresponding action of $\mathbb{T}^2$ on D_1 such that F is $\mathbb{T}^2$-equivariant.) The final question is offered as one way of asking whether or not the appearance of automorphisms is in some sense essential.

Question 3. *Can it happen that sing supp F is a closed complex manifold with discrete automorphism group?*

References.

[Ba1] D. Barrett, Regularity of the Bergman projection and local geometry of domains, Duke Math. J. **53** (1986), 333-343.

[Ba2] ---, Biholomorphic domains with inequivalent boundaries, Invent. Math. **85** (1986), 373-377.

[Be] S. Bell, Biholomorphic mappings and the $\bar{\partial}$-problem, Ann. Math. **114** (1981), 103-113.

[BBC] E. Bedford, S. Bell, and D. Catlin, Boundary behavior of proper holomorphic mappings, Mich. Math. J. 30 (1983), 107-111.

[C] D. Catlin, Subelliptic estimates for the $\bar{\partial}$-Neumann problem on pseudoconvex domains (to appear).

Compactness of families of holomorphic mappings up to the boundary

S. Bell
Purdue University
W. Lafayette, IN 47907

1. Introduction. David Catlin has shown ([11,12]) that the Bergman projection associated to a smooth bounded pseudoconvex domain of finite type (in the sense of D'Angelo [14]) satisfies strong pseudo-local estimates at each boundary point. Thus, Norberto Kerzman's proof [16] can be adapted to this class of domains and we are able to conclude that the Bergman kernel function associated to a smooth bounded pseudoconvex domain Ω of finite type extends C^∞ smoothly to $\bar{\Omega}\times\bar{\Omega}$ minus the boundary diagonal (see [6]). Recently, Harold Boas [9] and I [6] independently generalized Kerzman's theorem to a wider class of domains.

In most applications of Kerzman's theorem to the problem of boundary behavior of holomorphic mappings, only the fact that the Bergman kernel extends smoothly to $\bar{\Omega}\times\Omega$ is needed. I intend to demonstrate in this paper that the full statement of Kerzman's theorem has important consequences in the study of families of biholomorphic and proper holomorphic mappings.

Suppose that Ω_1 and Ω_2 are bounded pseudoconvex domains of finite type in $\mathbf{C}^n$ with C^∞ smooth boundaries and suppose that $\{f_i\}$ is a sequence of biholomorphic mappings $f_i : \Omega_1 \longrightarrow \Omega_2$. By passing to a subsequence, if necessary, we may assume that the f_i converge uniformly on compact subsets of Ω_1 to a holomorphic mapping $f:\Omega_1 \longrightarrow \bar{\Omega}_2$. It is a classical theorem of Cartan [11] (see [17], page 78) which states that f is either a biholomorphic mapping of Ω_1 onto Ω_2 or f is a mapping of Ω_1 into $b\Omega_2$, the boundary of Ω_2. I shall use the full result on the smooth extendibility of the

Bergman kernel to prove

THEOREM 1. **A)** In case f is biholomorphic, the components of the mappings f_i converge in $C^\infty(\bar{\Omega}_1)$ to the corresponding components of f. **B)** If f is a mapping of Ω_1 into $b\Omega_2$, and if the inverse mappings $F_i = f_i^{-1}$ converge uniformly on compact subsets of Ω_2 to a mapping F, then there is a point $p_1 \in b\Omega_1$ and a point $p_2 \in b\Omega_2$ such that the mappings f_i converge uniformly on compact subsets of $\bar{\Omega}_1 - \{p_1\}$ to the constant mapping $f \equiv p_2$.

It was observed by David Barrett [1] that Theorem 1, part A, follows as a consequence of the representation of the f_i in Bergman-Ligocka coordinates used in [8] and the fact that pseudoconvex domains of finite type satisfy condition R. I will give a new proof of this result which will generalize to the case where f is a map into the boundary. Barrett's proof of Theorem 1, part A, and the proof given here are valid in the more general setting where Ω_1 and Ω_2 are smooth bounded domains in $\mathbf{C}^n$ which satisfy condition R. (E. Bedford gave an alternate proof of this result in [3]. R. Greene and S. Krantz proved Theorem 1, part A, in [15] for strictly pseudoconvex domains.)

Theorem 1, part B, sounds new even in the strictly pseudoconvex case. However, if p_1 or p_2 is a strictly pseudoconvex boundary point, then both domains must be biholomorphic to the ball by Rosay's Theorem [18]. Hence, the theorem is only interesting, and only new, in case p_1 and p_2 are weakly pseudoconvex boundary points.

The hypothesis that the inverses converge in part B of Theorem 1 may seem strange. However, even in case Ω_1 and Ω_2 are both equal to the unit disc in $\mathbf{C}^1$, this hypothesis is necessary. Indeed, a typical sequence of automorphisms of the disc which converges to a boundary mapping is given by $f_k(z) = \exp(i\theta_k)(z - r_k\exp(i\psi_k))/(1 - zr_k\exp(-i\psi_k))$ where $\{r_k\}$ is a sequence of real numbers $0 \le r_k < 1$ such that r_k tends to one as k tends to infinity. This sequence converges uniformly on compact subsets of the disc to a boundary mapping if and only if $\theta_k+\psi_k$ converges modulo 2π. The convergence occurs up to the boundary away from a single point in the boundary if and only if we also have that ψ_k converges modulo 2π, and this happens if and only if the inverse mappings $F_k=f_k^{-1}$ converge uniformly on compact subsets of the disc. If the sequence $\{\psi_k\}$ forms a dense subset of the interval $[0,2\pi]$ modulo 2π, then the mappings f_k do not converge up to the boundary near any boundary point of the disc.

It should be mentioned that, throughout this paper, the hypothesis that Ω be smooth, bounded, pseudoconvex and of finite type can be replaced with the more general hypothesis that Ω be a smooth bounded domain whose Bergman projection satisfies weak pseudo-local estimates (in the sense of [6]) at each boundary point. Alternatively, the finite type hypothesis can be replaced by the hypothesis that Ω is a smooth, bounded pseudoconvex domain whose $\bar{\partial}$-Neumann operator is hypoelliptic.

The methods used in the proof of Theorem 1 will also yield

Theorem 2. _If_ $f_i : \Omega_1 \longrightarrow \Omega_2$ _is a sequence of biholomorphic mappings between bounded weakly pseudoconvex domains of finite type in_ $\mathbf{C}^n$ _with_ C^∞ _smooth boundaries which converges to a mapping_ f _into the boundary of_ Ω_2, _then, the mappings_ f_i _converge uniformly to a constant map on compact subsets of_ $\Omega_1 \cup \Gamma$ _where_ Γ _denotes the set of strictly pseudoconvex boundary points of_ Ω_1.

Note that there is no asssumption made about the convergence of the inverses in Theorem 2.

Actually, the proof of Theorem 1, part B, given below yields a stronger result than the statement indicates. Let the Bergman kernel function associated to Ω_1 be denoted by $K_1(z,w)$ and let $u_i = \det\,[f_i{}']$ denote the holomorphic jacobian determinant of f_i. With a little additional effort, I can prove

Theorem 3. If f is a mapping into $b\Omega_2$, and if the inverse mappings $F_i = f_i^{-1}$ converge then, in addition to the conclusions of Theorem 1, the sequence of functions u_i converges to the zero function in $C^\infty(\bar{\Omega}_1 - \{p_1\})$, and the mappings f_i converge to $f \equiv p_2$ in $C^\infty(\bar{\Omega}_1 - (S \cup \{p_1\}))$ where S is the set of points z in $b\Omega_1 - \{p_1\}$ where $K_1(z,p_1) = 0$.

Note that the set S in Theorem 3 is rather small because, according to [4] (see also [8]), the function $K_1(z,p_1)$ cannot vanish identically in z on Ω_1. Thus $K_1(z,p_1)$ cannot vanish on any open subset of $b\Omega_1$. The set S need not be empty as Harold Boas' counterexample to the Lu Qi-Keng conjecture [10] reveals.

With considerable additional effort, I can replace the set S in Theorem 3 by the smaller set consisting of points z in $b\Omega_1$ where $K_1(z,p_1)$ vanishes to infinite order in z. I will only sketch the proof of this result in section 2. (This smaller set must presumably be empty if Ω_1 is of finite type, although I have not been able to prove this.)

Versions of Theorems 1, 2 and 3 can be stated for proper holomorphic mappings. These theorems will be discussed in section 3 of this paper.

2. The proofs of Theorems 1, 2, and 3. Let $K_1(z,w)$ and $K_2(z,w)$ denote the Bergman kernel functions associated to Ω_1 and Ω_2, respectively. Because Ω_1 and Ω_2 are smoorh, bounded, pseudoconvex domains of finite type, $K_i(z,w)$ extends to be in $C^\infty(\bar{\Omega}_i \times \bar{\Omega}_i - \Delta_i)$ where $\Delta_i = \{(z,z) : z \in b\Omega_i\}$ for $i = 1,2$. Also, it has been shown by Catlin ([12] , [13]) that smooth, bounded, pseudoconvex domains of fintie type satisfy Condition R. Therefore, according to [4] (see also [8]), the complex linear span of the set Λ of functions $h(z)$ such that $h(z) = K_2(z,w)$ for some $w \in \Omega_2$ forms a dense subspace of $A^\infty(\Omega_2)$. (Here, $A^\infty(\Omega_2)$ denotes the subspace of $C^\infty(\bar{\Omega}_2)$ consisting of functions which are holomorphic on Ω_2.) The following fact is proved in [5] as a consequence of the denseness of the span of Λ in $A^\infty(\Omega_2)$.

Fact 1. Given a positive integer s, there exist finitely many points $w_1, w_2, \ldots, w_M$ in Ω_2 with the following property. Given any function h in $A^\infty(\Omega_2)$ and any point p in $\bar{\Omega}_2$, there exist constants $c_i = c_i(p)$ such that the function

$$\Gamma(z) = \sum_{i=1}^{M} c_i K_2(z, w_i)$$

agrees to order s with $h(z)$ at p. Furthermore, the constants c_i are uniformly bounded by $C \|h\|_s$ where

$$\|h\|_s = \sup \{ |(\partial^\alpha/\partial z^\alpha) h(z)| : z \in \bar{\Omega}_2, |\alpha| \le s \}$$

and C is a positive constant which does not depend on h or p.

I will prove this result here for the convenience of the reader. Fix a point p_0 in $\bar{\Omega}_2$. Let N be equal to the number of multi-indices α of length n with $0 \le |\alpha| \le s$ and let α_i, $i=1,\ldots,N$, be an enumeration of these multi-indices. Let D_i denote the differential operator $\partial^\alpha/\partial z^\alpha$ with $\alpha = \alpha_i$. I claim that there exist points $w_1, w_2, \ldots, w_N$ in Ω_2 such that the determinant of the matrix $[e_{ij}]$ where $e_{ij}(z) = D_i K_2(z, w_j)$ is non-zero for z near p_0. (Here, the operator D_i is acting in the z-variable only.) Indeed, if $\det [e_{ij}(p_0)] \equiv 0$ for all choices of $(w_1, w_2, \ldots, w_N)$ in $(\Omega_2)^N$, then, because of the multi-linear property of the determinant and the denseness of the span of Λ in $A^\infty(\Omega_2)$, it would follow that $\det[D_i g_j(p_0)] \equiv 0$ for all functions $g_1, g_2, \ldots, g_N$ in $A^\infty(\Omega_2)$. By letting $g_i(z) = (z - p_0)^\alpha$ with $\alpha = \alpha_i$, we obtain a contradiction. Let the constants $c_i = c_i(p_0)$ be given by the solution to the linear system

$$D_i h(p_0) = \sum_{j=1}^{N} c_j D_i K_2(p_0, w_j)$$

The constants c_i and the points w_i, so chosen, clearly satisfy the required properties at p_0. The estimate $|c_i(p)| \leq C \|h\|_s$ can be made uniform in p by enlarging the set of points $\{w_i\}$ if necessary because $\bar{\Omega}_2$ is compact and because $K_2(z,w)$ is C^∞ smooth on $\bar{\Omega}_2 \times \Omega_2$. It is interesting to note that the points in the finite set $\{w_i\}$ may be restricted to be in any open subset of Ω_2 because, as was shown in [5], the linear span of functions of the form $h(z) = K_2(z,w)$ is dense in $A^\infty(\Omega_2)$ for w restricted to any open subset of Ω_2. We will not need to know this more general result.

Let F_i denote the inverse mappings to the f_i and let $U_i = \det [F_i']$. The Bergman kernels transform according to

$$u_i(z) K_2(f_i(z),w) = K_1(z, F_i(w)) \overline{U_i(w)}. \tag{2.1}$$

Let $h(z)$ be any function in $A^\infty(\Omega_2)$ and let s be a positive integer. Fact 1 together with the transformation formula for the Bergman kernels implies that there exist finitely many points $\{w_j\}$ in Ω_2 with the following property. Given any point $p \in \Omega_1$, there exist constants $c_j = c_j(f_i(p))$ such that $h(z)$ agrees to order s with the function

$$\sum_{j=1}^{M} c_j K_2(z, w_j)$$

at $f_i(p)$. The constants c_j are uniformly bounded independent of the choice of p. Now the transformation formula (2.1) implies that the function $u_i(z) h(f_i(z))$ agrees to order s with the function

$$\Gamma_h(z) = \sum_{j=1}^{M} c_j K_1(z,F_i(w_j))\overline{U_i(w_j)} \qquad (2.2)$$

at $z = p$.

To prove Theorem 1, part A, note that the inverse mappings F_i converge uniformly on compact subsets of Ω_2 to a biholomorphic mapping $F : \Omega_2 \longrightarrow \Omega_1$. Of course $F = f^{-1}$. Thus, because the functions F_i and U_i converge uniformly on the compact set of points $\{w_j\}$, formula (2.2) reveals that the function $\Gamma_h(z)$ and its derivatives up to order s are uniformly bounded on Ω_1, independent of the choice of p. Hence, we are able to conclude that the functions $u_i(z)h(f_i(z))$ together with their derivatives up to order s are uniformly bounded on Ω_1. Now, because s was arbitrary and because the functions $u_i(z)h(f_i(z))$ converge uniformly on compact subsets of Ω_1 to $u(z)h(f(z))$, we are able to deduce that $u_i(z)h(f_i(z))$ converges in $A^\infty(\Omega_1)$ to $u(z)h(f(z))$. (Here $u(z) = \det [f']$.)

Let $h(z)\equiv 1$ to see that u_i converges in $A^\infty(\Omega_1)$ to u. Since $u\neq 0$ on $\bar{\Omega}_1$, we may now let $h(z)$ be equal to the j-th coordinate function z_j to deduce that $(f_i)_j$ converges in $A^\infty(\Omega_1)$ to $(f)_j$. This finishes the new proof of Theorem 1, part A. Note that we have not used the full hypotheses; we only needed to know that Ω_1 and Ω_2 were smooth bounded pseudoconvex domains which satisfy condition R.

We shall use a similar argument to prove Theorem 1, part B. First, note that the inverse mappings F_i also converge uniformly on compact subsets of Ω_2 to a boundary mapping $F : \Omega_2 \longrightarrow b\Omega_1$.

We will now prove that there is a point $p_2 \in b\Omega_2$ such that $f_i(z)$ tends to p_2 uniformly on compact subsets of Ω_1. To see this, note that $u_i(z)$ converges uniformly on compact subsets of Ω_1 to the zero function as i tends to infinity according to

Cartan's theorem [11] (see also [17], page 78, Theorem 4). The transformation formula for the kernel functions can be written

$$u_i(z)\, K_2(f_i(z),f_i(w))\, \overline{u_i(w)} = K_1(z,w)\,. \tag{2.3}$$

Pick points z and w in Ω_1 such that $K_1(z,w) \neq 0$. Since $u_i(z)$ and $u_i(w)$ tend to zero, we deduce from formula (2.3) that $K_2(f_i(z),f_i(w))$ becomes large in modulus as i tends to infinity. The smoothness of the Bergman kernel away from the boundary diagonal now forces us to conclude that $f_i(z)$ and $f_i(w)$ tend to the same point in $b\Omega_2$. Suppose z_1 and z_2 are two points in Ω_1. Let $w \in \Omega_1$ be such that $K_1(z_1,w) \neq 0$ and $K_1(z_2,w) \neq 0$. The above argument yields that $f_i(z_1)$ and $f_i(z_2)$ both converge to the same point in $b\Omega_2$; call it p_2. (It should be remarked that this fact could have been deduced from the finite type condition on Ω_2 because a domain of finite type cannot contain a non-trivial holomorphic curve in its boundary [14]). Similarly, it can be shown that there is a point $p_1 \in b\Omega_1$ such that $F_i(w)$ tends to p_1 uniformly on compact subsets of Ω_2.

Let $h(z) \equiv 1$. Arguing as in the proof of Theorem 1, part A, we assert that given a positive integer s, there exist finitely many points $\{w_j\}$ in Ω_2 with the following property. Given any point $z \in \Omega_1$, there exist constants $c_j = c_j(f_i(z))$ such that $u_i(z) = u_i(z)h(f_i(z))$ agrees to order s with the function

$$\Gamma_h(z) = \sum_{j=1}^{M} c_j\, K_1(z,F_i(w_j))\, \overline{U_i(w_j)} \tag{2.4}$$

at z. The constants c_j are uniformly bounded independent of the choice of z. Now, because s was arbitrary and because $F_i(w_j) \longrightarrow p_1$ and $U_i(w_j) \longrightarrow 0$, formula (2.4) and the differentiability of $K_2(z,w)$ on $(\bar{\Omega}_2 \times \bar{\Omega}_2) - \Delta_2$ imply that u_i tends to the zero function in $C^\infty(\bar{\Omega}_1 - \{p_1\})$.

We shall now use the transformation formula in the form (2.3) to prove that f_i tends to the constant mapping p_2 uniformly on any compact subset X of $\bar{\Omega}_1 - \{p_1\}$. Let z_0 be a point in X and let w be a point in Ω_1 such that $K_1(z_0,w) \neq 0$. That such a point w exists is an easy consequence of the denseness of the linear span of $\{K_1(z,w): w \in \Omega_1\}$ in $A^\infty(\Omega_1)$ (see also [4]). Let V be an open neighborhood of z_0 in X such that there exists a positive constant c with $|K_1(z,w)| > c$ for all $z \in V$. Since $u_i(z)$ tends to zero uniformly for $z \in V$, and since $u_i(w)$ tends to zero, the transformation formula (2.3) yields that $K_2(f_i(z),f_i(w))$ becomes uniformly large in modulus for $z \in V$ as i tends to inifinity. Thus, $f_i(z)$ becomes uniformly close to the boundary point p_2 to which $f_i(w)$ converges. This finishes the proof of Theorem 1, part B.

The proof of Theorem 2 is accomplished by observing that, by Rosay's Theorem [18], if z is a strictly pseudoconvex boundary point of Ω_1, then for each compact subet K of Ω_2, there exists a ball $B(z;\delta)$ about z in $\mathbf{C}^n$ such that $F_i(K) \cap B(z;\delta) = \phi$ for all i. Here, we have used the fact that because the domains are weakly pseudoconvex, they cannot be biholomorphic to the ball. It will now be shown that U_i tends to zero uniformly on compact subsets of Ω_2 even if the sequence F_i does not converge. Indeed, since $|U_i|^2$ is equal to the real jacobian determinant of F_i when viewed as a mapping from $\mathbf{R}^n$ into itself, it follows that the integral of $|U_i|^2$ over a compact subset $\mathbf{K}$ of Ω_2 is equal to the volume of $F_i(\mathbf{K})$. Since f_i converges to a boundary mapping, the set $F_i(\mathbf{K})$ is contained in a small neighborhood of $b\Omega_1$ for sufficiently large i. Thus, the volume of $F_i(\mathbf{K})$ is small, and we deduce that $U_i(w)$ tends to zero uniformly on compact subsets of Ω_2. It is now an easy matter to adapt the proof of

Theorem 1, part B, given above to the setting of Theorem 2.

To prove Theorem 3, note that we have already shown that u_i. tends to the zero function in $C^\infty(\bar{\Omega}_1 - \{p_1\})$. To finish the proof, we must verify the following claim.

Claim 1. If z_1 and z_2 are two points in $\bar{\Omega}_1 - \{p_1\}$ and if $K_1(z_2,p_1) \neq 0$, then the ratio $u_i(z_1)/u_i(z_2)$ tends to a finite limit as i tends to infinity.

To prove the claim, note that the transformation formula (2.1) implies that

$$\frac{u_i(z_1)\, K_2(f_i(z_1),w)}{u_i(z_2)\, K_2(f_i(z_2),w)} = \frac{K_1(z_1,F_i(w))}{K_1(z_2,F_i(w))} \tag{2.5}$$

for any point w in Ω_2. Choose w so that $K_2(p_2,w) \neq 0$. Then, since $f_i(z_1)$ and $f_i(z_2)$ tend to p_2 and $F_i(w)$ tends to p_1, we obtain that the ratio $u_i(z_1)/u_i(z_2)$ tends to the constant $K_1(z_1,p_1)/K_1(z_2,p_1)$. The claim is proved.

Pick a point b in Ω_2 such that $K_2(p_2,b) \neq 0$. Define $\Phi_i(z) = u_i(z)/\overline{U_i(b)}$. I now claim that Φ_i converges in $C^\infty(\bar{\Omega}_1 - \{p_1\})$ to the function $\Phi(z) = K_1(z,p_1)/K_2(p_2,b)$. Indeed, that Φ_i converges uniformly on compact subsets of $\bar{\Omega}_1 - \{p_1\}$ to Φ follows imediately from the transformation formula in the form (2.1) and the convergence properties of f_i and F_i that we have already established. Let $h(z)$ be any function in $A^\infty(\Omega_2)$. I will now prove that $\Phi_i(z)h(f_i(z))$ converges in $C^\infty(\bar{\Omega}_1 - \{p_1\})$ to $\Phi(z)h(p_2)$. As before, let s be a positive integer and let $\{w_j\}$ be a finite set of points in Ω_2 with the property that given any point $z \in \Omega_1$, there exist constants $c_j = c_j(f_i(z))$ such that $\Phi_i(z)\, h(f_i(z))$ agrees to order s with the function

$$\Gamma_h(z) = \sum_{j=1}^{M} c_j\, K_1(z, F_i(w_j))\, \overline{U_i(w_j)/U_i(b)} \tag{2.6}$$

at z. The constants c_j are uniformly bounded independent of the choice of z. Claim I applied to the inverse mappings F_i now yields that the numbers $U_i(w_j)/U_i(b)$ tend to a finite limit as i tends to infinity. Thus we are able to conclude, by virtue of formula (2.6), that the functions $\Phi_i(z)h(f_i(z))$ together with their derivatives up to order s are uniformly bounded on compact subsets of $\bar{\Omega}_1 - \{p_1\}$. Since $\Phi_i(z)h(f_i(z))$ converges uniformly on compact subsets of $\bar{\Omega}_1 - \{p_1\}$ and because s was arbitrary, we conclude that the convergence must take place in $C^\infty(\bar{\Omega}_1 - \{p_1\})$. Note that the set S in the statement of Theorem 3 is precisely the set of points z in $b\Omega_1 - \{p_1\}$ where $\Phi(z) \neq 0$. Finally, by letting $h(z) = z_j$, the j-th coordinate function, we deduce that the mappings f_i converge in $C^\infty(\bar{\Omega}_1 - (S \cup \{p_1\}))$ to the constant map p_2. This finishes the proof of Theorem 3.

Remark. To prove the stronger theorem mentioned after the statement of Theorem 3 in section 1, a division problem with uniform bounds must be solved. We know that $\Phi_i(z)h(f_i(z))$ tends to $\Phi(z)h(p_2)$ in $C^\infty(\bar{\Omega}_1 - \{p_1\})$ for each function h in $A^\infty(\Omega_2)$, in particular for $h = z^\alpha$. The division theorem of [7] can be proved with uniform bounds at any point where $\Phi(z)$ vanishes to finite order to yield that therefore $f_i(z)$ tends to p_2 in C^∞ of the larger set.

3. The Proper Mapping Case. In this section, we will only consider families $\{f_i\}$ of proper holomorphic mappings $f_i : \Omega_1 \longrightarrow \Omega_2$ between bounded pseudoconvex domains in C^n with real analytic boundaries. No proofs of any of the assertions made in this section will be given.

If Ω_1 and Ω_2 are pseudoconvex domains with real analytic boundaries in $\mathbb{C}^n$, then it follows from results in [2] that there are only finitely many possible proper holomorphic mappings $P_m : \Omega_1 \longrightarrow \Omega_2$, $m=1,2,...,M$, in the sense that given a proper holomorphic mapping $f:\Omega_1 \longrightarrow \Omega_2$, there exists an automorphism Φ of Ω_2 such that $f = \Phi \circ P_m$ for some m, $1 \le m \le M$. Hence, if $\{f_i\}$ is a sequence of proper holomorphic mappings between Ω_1 and Ω_2, then by passing to a subsequence, if necessary, we may suppose that there is a sequence of automorphisms $\{\Phi_i\}$ of Ω_2 which converge to a holomorphic mapping $\Phi : \Omega_1 \longrightarrow \bar{\Omega}_2$ and a proper holomorphic mapping $P : \Omega_1 \longrightarrow \Omega_2$ such that $f_i = \Phi_i \circ P$. Thus, the study of the convergence properties of the sequence $\{f_i\}$ is reduced to Theorems 1, 2 and 3. We obtain

Theorem 4. If $\{f_i\}$ is a sequence of proper holomorphic mappings $f_i : \Omega_1 \longrightarrow \Omega_2$ between pseudoconvex domains in $\mathbb{C}^n$ with real analytic boundaries which converges uniformly on compact subsets of Ω_1 to a mapping $f : \Omega_1 \longrightarrow \mathbb{C}^n$, then either

A) there exist a proper holomorphic mapping $P : \Omega_1 \longrightarrow \Omega_2$ and a positive integer N such that for $i > N$, there is a sequence of automorphisms $\{\Phi_i\}$ of Ω_2 which converge in $C^\infty(\bar{\Omega}_2)$ to an automorphism Φ of Ω_2 with the property that $f_i = \Phi_i \circ P$ for $i > N$. Thus, the sequence $\{f_i\}$ converges in $C^\infty(\bar{\Omega}_1)$ to the proper mapping $\Phi \circ P$. Or

B) by passing to a subsequence, we may assume that there is a finite set of points $\{p_1, p_2, ..., p_k\} \subset b\Omega_1$ and a point $q \in b\Omega_2$ such that f_i tends to q uniformly on compact subsets of $\bar{\Omega}_1 - \{p_1, ..., p_k\}$.

It is possible to prove versions of Theorem 4 in the finite type setting by applying the transformation formula for the Bergman kernel function under proper holomorphic mappings (see [5]) and arguing as in the proof of Theorems 1, 2, and 3 above. The details of this argument will appear elsewhere.

References

1. D. E. Barrett, **Regularity of the Bergman projection on domains with transverse symmetries**, Math. Ann.**258** (1982), 441-446.
2. E. Bedford, **Proper holomorphic mappings from domains with real analytic boundary**, Amer. J. Math. **106** (1984), 745-760.
3. E. Bedford, **Action of the automorphisms of a smooth domain in C^n**, Proc. A. M. S. **93** (1985), 232-234.
4. S. Bell, **Non-vanishing of the Bergman kernel function at boundary points of certain domains in C^n**, Math. Ann. **244** (1979), 69-74.
5. S. Bell, **Boundary behavior of proper holomorphic mappings between non-pseudoconvex domains**, Amer. J. Math. **106** (1984), 639-643.
6. S. Bell, **Differentiability of the Bergman kernel and pseudo-local estimates**, Math. Zeit., in press.
7. S. Bell and D. Catlin, **Boundary regularity of proper holomorphic mappings**, Duke Math. J. **49** (1982), 385-396.
8. S. Bell and E. Ligocka, **A simplicfication and extension of Fefferman's theorem on biholomorphic mappings**, Invent. Math. **57** (1980), 283-289.
9. H. P. Boas, **Extension of Kerzman's theorem on differentiability of the Bergman kernel function**, to appear.
10. H. P. Boas, **Counterexample to the Lu Qi-Keng conjecture**, to appear.
11. H. Cartan, **Sur les fonctions de plusieurs variables complexes: L'itération des transformations intérieurs d'un domaine borné**, Math. Zeit. **35** (1932), 760-773.
12. D. Catlin, **Boundary invariants of pseudoconvex domains**, Ann. Math. **120** (1984) 529-586.
13. D. Catlin, **Subelliptic estimates for the $\bar\partial$-Neumann problem on pseudoconvex domains**, Ann. Math., to appear.
14. J. P. D'Angelo, **Real hypersurfaces, orders of contact, and applications**, Ann. Math.**115** (1982), 615-637.
15. R. Greene and S. Krantz, **The automorphism groups of strongly pseudoconvex domains**, Math. Ann. **261** (1982), 425-446.
16. N. Kerzman, **The Bergman kernel function. Differentiability at the boundary**, Math.

Ann. **195** (1972), 149-158.

17. R. Narasimhan, Several Complex Variables, Chicago Lectures in Mathematics, University of Chicago Press, 1971.

18. J. P. Rosay, **Sur une caractérisation de la boule parmi les domaines de C^n par son groupe d'automorphismes,** Ann. Inst. Fourier **29** (4) (1979), 91-97.

The Imbedding Problem For Open Complex Manifolds

J.S. Bland

A problem of considerable interest in the study of several complex variables is to characterize which open complex manifolds are biholomorphically equivalent to open complex submanifolds of complex Euclidean space, and further, to characterize those which are equivalent to bounded open submanifolds. This problem is not only important in that it helps to delineate how bad higher dimensional complex manifolds can be (compare with the uniformization theorem of one complex variable), but it also has an intrinsic beauty in that it interrelates the various mathematical properties of the manifold. In the case that one is interested in proper imbeddings into C^N ($N \geq n$, where n is the dimension of the manifold), it is well known that a necessary and sufficient condition for the existence of such an imbedding is the existence of a smooth strongly pseudoconvex exhaustion function. More recently, this function theoretic property has been related to the existence of complete Kähler metrics with certain prescribed curvature properties (see section 1). In the case of bounded domains, it has long been recognized that various geometric properties of the boundary of the domain are related to the function theoretic properties of the domain. More recently, it has been recognized that this relationship can be extended to a certain class of open complex manifolds by associating to such manifolds a natural boundary, and that both the intrinsic function theoretic properties of the domain and the extrinsic geometric properties of the boundary can be related to the existence of complete Kähler metrics with certain additional prescribed properties. The main concern of this paper is to outline some of the results in these directions, and to indicate the various techniques needed to understand such manifolds.

§ 1: Stein Manifolds

Stein manifolds form a broad yet cohesive class of open complex manifolds. The cohesiveness of the class is amply indicated by the two following well-known characterizations: An open complex manifold M^n is Stein iff it admits a smooth strongly pseudoconvex exhaustion function iff it admits a smooth proper holomorphic imbedding into C^{2n+1}, [5, 25, 44, 45]. The breadth of the class is illustrated by the fact that in the uniformization theorem of one complex variable, both the unit disc and the complex plane are Stein. In this very fact, however, lies one of the major difficulties with this class of manifolds: it suppresses all information concerning the existence of bounded holomorphic functions, and the vastly different function theoretic properties of the unit disc and the complex plane.

This section will concentrate upon geometric properties that guarantee that an open complex manifold is Stein, as well as more stringent geometric properties which will guarantee respectively the non existence of any bounded holomorphic functions and biholomorphic equivalence to $\mathbf{C}^n$. The remaining sections will study properties specific to bounded Stein submanifolds of $\mathbf{C}^N$ ($N \geq n$).

In [26, 28, 29, 30], Greene and Wu determined several alternative formulations of sufficient differential geometric conditions to guarantee Steinness. They stated the following:

(1.1) Theorem. Let M be a complete Kähler manifold. Then M is Stein if any of the following holds:

(a) M is simply connected with nonpositive sectional curvature

(b) M is noncompact with nonnegative sectional curvature, and moreover, the sectional curvature is strictly positive outside a compact subset of M.

(c) M is noncompact with nonnegative sectional curvature, and the holomorphic bisectional curvature is strictly positive

(d) M is noncompact with strictly positive Ricci curvature, nonnegative sectional curvature, and the canonical bundle is trivial.

Notice that more can be said about the manifold in parts (b), (c) and (d). In each case, the manifold is Stein and cannot admit any bounded holomorphic functions; this latter fact follows from estimates by Yau [49] for harmonic functions on complete Riemannian manifolds with nonnegative Ricci curvature. This observation also indicates one strong motivation for working with Kähler metrics: on Kähler manifolds, holomorphic functions are automatically harmonic.

The main technique in establishing this theorem is to use the differential geometric properties to establish the existence of a strictly plurisubharmonic exhaustion function. The various curvature assumptions on complete Riemannian manifolds are used to estimate the convexity of the distance function and functions derived from the distance function. Since the metric is Kähler, the geometric properties are closely tied to the complex structure of the manifold; in particular, smooth strongly convex functions are strongly pseudoconvex (see [27]).

Siu and Yau used differential geometric properties in an elegant fashion in [47] to characterize manifolds which are biholomorphically equivalent to $\mathbf{C}^n$. They showed that if M is a complete open simply connected Kähler manifold with $-A/r^{2+\varepsilon} \leq$ sectional curvature ≤ 0 for some positive constants A, ε (r is distance from a fixed point p), then M is biholomorphic to $\mathbf{C}^n$. Their procedure to prove this theorem was roughly as follows. They first solved the $\bar{\partial}$ problem on (n,1) forms in suitably chosen weighted L_2 spaces. By cleverly choosing the weights, they were able to establish the existence of a holomorphic n form which was nonzero at p and satisfied certain growth estimates at infinity; moreover they were able to show that any two holomorphic n forms satisfying these same growth properties were constant multiples of each other. They then changed the weight factors and solved another $\bar{\partial}$ problem on (n,1) forms to produce a holomorphic n form which vanished at p; the ratio of the new holomorphic n form to the first one was a holomorphic function vanishing at p and of "minimal" growth. This function acted like a linear holomorphic function on $\mathbf{C}^n$. They found n such functions forming a local coordinate system at p, and then showed that these functions indeed provided the components for a biholomorphic map from M to $\mathbf{C}^n$.

Mok, Siu and Yau weakened the geometric hypotheses needed to guarantee biholomorphic equivalence to $\mathbf{C}^n$. They established the following result [42]:

(1.2) Theorem. Let M^n be an n complex dimensional complete Kähler manifold ($n \geq 2$) with a pole p (i.e. - the exponential map centred at p is a diffeomorphism). Suppose that the norm of the sectional curvature is $\leq A/(1 + r^2)^{1+\varepsilon}$ for some constants ε, and A depending on ε. Then $M \cong \mathbf{C}^n$. If, in addition, M has nonpositive sectional curvature, then M is isometrically biholomorphic to $\mathbf{C}^n$.

Notice that this theorem not only gives intrinsic geometric characterizations of open complex manifolds which are biholomorphically equivalent to $\mathbf{C}^n$, but it also establishes an important gap theorem for complete Kähler manifolds with nonpositive sectional curvature: if the curvature dies too quickly at ∞, then the metric is identically flat. This latter phenomenon is really a differential geometric phenomenon; similar results hold for complete Riemannian manifolds [31].

§ 2: Geometry of Bounded Strongly Pseudoconvex Domains

Bounded domains of holomorphy form an important subclass of all Stein manifolds. Their distinctive character arises from the fact that bounded domains admit lots of bounded holomorphic functions. (Compare, for example, the function theory of the unit disc with the function theory of the complex plane in one complex variable). Further, results of Hartogs, Levi and Oka lead one to consider the relationship between certain geometric properties of the boundary (pseudoconvexity) and the analysis on the domain. In particular, from knowing certain aspects of the function theory of the domain, one can deduce certain geometric properties of the boundary. Our approach to the study of bounded domains of holomorphy will be as follows. We will start with bounded strongly pseudoconvex domains, and observe that not merely the property of pseudoconvexity, but the entire CR structure of the boundary (that is - the restriction of the complex structure to the boundary) is invariant under biholomorphic maps. The CR structure on the boundary can be studied intrinsically, and in the analytic case, there exist a complete list of invariants. These invariants can in principle be determined from the asymptotics of complete metrics on the domain. Finally, it will be observed that the boundary with its smooth CR structure can be fully determined from the asymtotics of complete Kähler metrics on the interior: that is, open complex manifolds admitting complete Kähler metrics with certain asymptotic properties admit a natural compactification. The payoff for this approach will be in § 3 when we study the function theory of such open complex manifolds by alternatively considering them as the interior of compact complex manifolds with smooth boundary.

C. Fefferman established the following result in [21]: If $F : D_1 \to D_2$ is a biholomorphic map between two smooth bounded strongly pseudoconvex domains in $\mathbf{C}^n$, then F extends smoothly to a diffeomorphism $F : \overline{D}_1 \to \overline{D}_2$. His proof involved two major steps. The first step was a deep analysis of the Bergman kernel on smooth bounded strongly pseudoconvex domains, and its singularity at the boundary. The second step used this information in an analysis of the geodesics for the associated complete Bergman metric; in this step, he showed that the geodesics, as curves in $\mathbf{C}^n$, extended smoothly to the boundary of the domain. Then, since biholomorphisms are isometries of the Bergman metric, the geodesics can be used to define biholomorphically invariant coordinate systems which are smooth to the boundary, and the map F must also be smooth to the boundary.

An immediate consequence of Fefferman's result is the following: If M is an open complex manifold which is biholomorphically equivalent to a smooth bounded strongly pseudoconvex domain $D \subset \mathbf{C}^n$, then there is naturally associated to M a smooth invariantly defined boundary, and the complex structure on M extends smoothly to a strongly pseudoconvex CR structure on the boundary. Indeed, for a fixed biholomorphism $M \cong D \subset\subset \mathbf{C}^n$, define ∂M by $M \cup \partial M \cong \overline{D}$. That this definition of the boundary is invariant follows from the fact that if $M \cong D'$ for any other smooth bounded strongly pseudoconvex domain D', then $D \cong M \cong D'$, and by Fefferman's result $\overline{D} \cong \overline{D'}$, where the last equivalence is a biholomorphism on the interior and a diffeomorphism on the closure.

At about the same time as Fefferman's results, Chern and Moser [17] studied the boundaries of smooth strongly pseudoconvex domains and were able to describe a complete list of local invariants. Moser's approach was extrinsic, describing a normal form for the boundary. Chern's approach was intrinsic, and generalizes immediately to abstract nondegenerate CR manifolds.

Let N be a nondegenerate CR manifold and θ a nonvanishing real one form which annihilates the holomorphic tangent space to N. Let E be the line bundle of positive multiples of θ. Chern showed that associated to E is a principal bundle Y (a subbundle of the bundle of coframes of E) with an invariantly defined (up to CR equivalence) Cartan connection with values in $\mathfrak{su}(n,1)$. Thus, in the real analytic case, a complete list of local invariants can be given in terms of the local invariants of the Cartan connection.

In [34], Klembeck began studying the asymptotics of complete Kähler metrics on bounded strongly pseudoconvex domains in $\mathbf{C}^n$. He showed that for a wide class of complete Kähler metrics, the holomorphic sectional curvature of the metric tends to a negative constant near the boundary. Choose a defining function ϕ for D which is smooth and strongly pseudoconvex on $\overline{D}$. Define a Kähler metric $g_{i\bar{j}}dz^i \otimes dz^{\bar{j}}$ by $g_{i\bar{j}} = -\partial^2\log(-\phi)/\partial z_i \partial z_{\bar{j}}$. (It follows from Fefferman's analysis of the Bergman kernel that the Bergman metric satisfies these estimates for the metric and its curvature). Then $g_{i\bar{j}} = \phi_{i\bar{j}}/(-\phi) + \phi_i\phi_{\bar{j}}/\phi^2$ is a complete Kähler metric which blows up at the rate $1/(-\phi)^{1/2}$ in complex tangential directions and $1/(-\phi)$ in complex normal directions. Klembeck's result follows easily by explicitly computing the curvature tensor for the metric and the tensor corresponding to constant negative holomorphic sectional curvature and comparing the highest order terms. This computation also makes it clear that estimates similar to the growth estimates for the metric hold for all higher order covariant derivatives of the curvature. This implies that the growth for any invariant tensor associated to the metric is well controlled on strongly pseudoconvex domains.

In [16], Cheng and Yau showed that on any smoothly bounded pseudoconvex domain in $\mathbf{C}^n$, there exists a unique complete Einstein- Kähler metric. Further, if the domain is strongly pseudoconvex, then the potential for the metric can be written as $g = -\log(-\phi)$ where ϕ is a smooth defining for function for D which is in $C^{\omega}(D) \cap C^{n+1}(\overline{D})$. It then follows from Klembeck's calculations that the holomorphic sectional curvature for this invariantly defined metric approaches -1 at infinity.

Fefferman took up the study of boundary invariants in [22]. In this paper, he started with a smooth strongly pseudoconvex defining function ϕ for D, and algebraically manipulated it to obtain a new function (which we will still call ϕ) which satisfied the equation $(-\phi + |d\phi|^2)\,|\phi_{i\bar{j}}| = 1 + 0(\phi)^{n+1}$. (In this notation, $\phi_{i\bar{j}} = \partial^2\phi/\partial z_i \partial z_{\bar{j}}$, $\phi_{i\bar{j}}\phi^{\bar{j}k} = \delta_i^k$, $|d\phi|^2 = \phi_i \phi_{\bar{j}} \phi^{i\bar{j}}$). Solutions to this equation automatically satisfy the requirement that (-log $(-\phi)$) is the potential for a complete Kähler metric which is asymptotically Einstein to a certain order. Then, using coordinates (z,w) on $D \times \mathbf{C}^*$, the function $u = \phi\,|w|^{2/n+1}$ is the potential for a nondegenerate hermitian metric which restricts to a circle bundle over ∂D as a nondegenerate Lorentz metric. The conformal class of this metric is biholomorphically invariant in the following sense: if the same procedure is applied to any biholomorphically equivalent strongly pseudoconvex domain, the resulting Lorentz metric can be pulled back by a biholomorphism of $D_1 \times \mathbf{C}^* \cong D_2 \times \mathbf{C}^*$ (which is induced by the biholomorphism $D_1 \cong D_2$) in such a way that the new Lorentz metric on the circle bundle is in the same conformal class as the original. Thus, this procedure results in an invariant conformal class associated to a circle bundle over the boundary.

Webster [48] used a simple construction to tie together Fefferman's approach to boundary invariants with Chern's approach. An alternative description of the relationship between the two approaches was described by Burns, Diederich and Schnider [11]. Their method had the additional advantage of providing an intrinsic derivation of Fefferman's conformal structure on the circle bundle. We will follow Webster's approach here because it is more closely linked with the extrinsic geometry of the boundary, and the asymptotics of complete approximate Einstein-Kähler metrics.

Webster initially followed Fefferman's approach, beginning with a smooth strongly pseudoconvex defining function ϕ and defining the potential for a smooth Kähler metric on $\overline{D} \otimes \mathbf{C}^*$ by $u = \phi |w|^{2/(n+1)}$ as above. He next reduced the bundle of frames on $\partial D \times \mathbf{C}^*$ to a principal bundle P of frames which are adapted to the boundary (adapted hemitian frames), and further to a subbundle P, of unitary adapted frames with a structure group K. Recall now Chern's construction: over the line bundle E (of positive multiples of the annihilator of the complex tangent space), he constructed a principal bundle $B_1^*(=Y)$ of coframes. Denote by B_1 the dual bundle of frames with structure group G, and let B denote the larger bundle of adapted frames (not necessarily unitary relative to a hermitian form). Webster exhibits a map f: $\partial D \times \mathbf{C}^* \rightarrow E$ which is the identity on the base ∂D, and associates to it a natural bundle map $\hat{f}$:P $\rightarrow$ B which makes P into an S^1 bundle over B, and induces an isomorphism from G to K. He can then pull back the Cartan connection on B, which was described by Chern, and compare it with the hermitian connection on P_1 associated to Fefferman's metric. This procedure can be carried out when Fefferman's metric is computed from an arbitrary strongly pseudoconvex defining function ϕ. When one assumes in addition that ϕ satisfies $(-\phi + |d\phi|^2)|\phi_{i\bar{j}}| = 1 + 0(\phi^3)$, then the curvature form for the Cartan connection pulls back to the curvature form for Fefferman's hermitian connection; moreover, the Cartan connection pulls back to the hermitian connection for some metric in the same conformal class as Fefferman's metric. All invariants from Chern's connection can be expressed in terms of invariants for the conformal class of metrics described by Fefferman. A byproduct of this result is that the intrinsic invariants of the boundary can be expressed in terms of the asymptotics of the complete Einstein Kähler metric on D; however, the methodology is extrinsic in nature.

A converse to Klembeck's result on the asymptotic behaviour of metrics on smooth bounded strongly pseudoconvex domains was established in [6]. In this paper, it was shown that if M is an open complex manifold with a complete Kähler metric g, and if for some point $p \in M$ and some sector of the exponential coordinate system centred at p, the exponential map is injective, and the curvature and its covariant derivatives satisfy the growth estimates of Klembeck's metrics (expressed intrinsically in terms of distance from the point p), then there exists a natural boundary at infinity for this sector and the complex structure on the coordinate system extends smoothly to a strongly pseudoconvex CR structure on the boundary. This boundary is natural in the following sense: if q is any other point, and some sector of the exponential coordinate system centred at q is injective and intersects the sector from p, the overlap map is smooth out to the closure of the coordinate systems. Hence, the boundary of M and its CR structure can be defined independent of the base point p [8]. A byproduct of this result is that the CR structure on the boundary, and its Chern invariants can be expressed intrinsically in terms of the asymptotics of an approximate Einstein Kähler metric on the interior of the manifold. The next section will use the compactification of open complex manifolds to study the function theory on the interior.

§ 3: Function Theory on Bounded Strongly Pseudoconvex Domains.

One natural approach to the function theory on open complex manifolds is via the $\bar\partial$-problem. In this approach, one starts with a smooth function with prescribed properties and modifies it by subtracting off the solution to an associated $\bar\partial$-problem. The net effect is to take the orthogonal projection of the original function onto the space of holomorphic functions in some Hilbert space of square integrable functions relative to a weight function. The Hilbert space may be changed according to the desired properties of the resultant functions. Siu and Yau applied this approach effectively to (n,0) forms in [47] to produce functions of minimal growth; they then showed that such functions played the role of linear functions, and used them to define a biholomorphic map to $\mathbf{C}^n$ (see § 1). However, no one has used this method successfully in producing bounded holomorphic functions.

On the other hand, Kohn [23, 35] has extensively studied the problem of solving the $\bar\partial$ equation on open complex manifolds M when M is a priori given as the interior of a compact complex manifold $\overline{M}$ with a smooth strongly pseudoconvex boundary. In his approach, he fixes a smooth hermitian metric on $\overline{M}$, and considers solutions to an associated elliptic equation. The formalism leads to a noncoercive boundary value problem, which makes the estimates for smoothness at the boundary difficult to obtain. However, once the estimates at the boundary are obtained, then the solutions to the equation are smooth on $\overline{M}$ if the initial data is smooth on $\overline{M}$, and in particular, the solutions are bounded.

In [23,35], Kohn solved the $\bar\partial$ Neumann problem in much greater generality than we presently need. In order to simplify the exposition, we shall only present the statement for complex manifolds with strongly pseudoconvex boundary. The manifold $\overline{M}$ will have a hermitian metric h. The square integrable forms on $\overline{M}$ form a Hilbert space, and $\bar{\partial}^*$ will be the Hilbert space adjoint of $\bar{\partial}$. We define the self adjoint elliptic operator $\Box = \bar{\partial}\bar{\partial}^* + \bar{\partial}^*\bar{\partial}$. Let $H^{p,q}(\overline{M})$ be the square integrable (p,q) forms on $\overline{M}$ which are in the kernel of $\Box$ and let $\Lambda^{p,q}(\overline{M})$ be the smooth (p,q) forms on $\overline{M}$. Kohn establishes the following:

(3.1) Theorem. Let $\overline{M}$ be a compact complex manifold with a smooth strongly pseudoconvex boundary and a hermitian metric h. Then

1) For $q > 0$, $H^{p,q}(\overline{M})$ is finite dimensional
2) $H^{p,q}(\overline{M}) \subset \Lambda^{p,q}(\overline{M})$
3) If $\alpha \in \Lambda^{0,1}(\overline{M})$, $\bar{\partial}\alpha = 0$ and α is perpendicular to $H^{0,1}(\overline{M})$, then there exists a unique $u \in \Lambda^{0,0}(\overline{M})$ such that $\bar{\partial}u = \alpha$ and satisfying the Sobolev space estimates $\|u\|_{s+1} \leq C\,\|\alpha\|_s$ for some constant C independent of α.

H. Rossi and J. Taylor referred to a compact complex manifold with a smooth strongly pseudoconvex boundary as a smooth finite strongly pseudoconvex manifold. (It should be noted that in their definition, they required only that the integrable complex structure on M extend smoothly to the closure $\overline{M}$. Thus, while the Newlander-Nirenberg theorem guarantees the existence of local holomorphic coordinates at interior points, there is no a priori assumption made on the existence of such coordinates at boundary points.) Such a manifold is said to be complete if, in addition, it admits a global smooth strongly pseudoconvex function. In [46], Rossi and Taylor showed how Kohn's results could be used to prove an imbedding theorem for smooth complete finite strongly pseudoconvex manifolds.

(3.2) Theorem. Let $\overline{M}$ be a smooth complete finite strongly pseudoconvex manifold. Then there exist holomorphic functions $f_1, \ldots, f_N \in \Lambda^{0,0}(\overline{M})$ such that $(f_1,\ldots,f_N)\colon \overline{M} \to \mathbb{C}^N$ is an imbedding.

Their proof proceeded as follows. First, the global strongly pseudoconvex function and the technique of Hörmander's weighted L^2 spaces can be used to show that $\dim(H^{0,1}(\overline{M})) = 0$. Thus, Kohn's theorem states that if $\alpha \in \Lambda^{0,1}(\overline{M})$ and $\bar{\partial}\alpha = 0$, then there exists $u \in C^\infty(\overline{M})$ such that $\bar{\partial}u = \alpha$ and $\|u\|_{s+1} \leq C\|\alpha\|_s$ for some constant C depending on s,$\overline{M}$ but not on α. Rossi and Taylor then used the same technique which Boutet de Monvel [10] used to imbed compact strongly pseudoconvex CR manifolds. Applying the technique in this situation and using Kohn's estimates, they were able to establish the existence of local coordinate systems for all boundary points;

more precisely, for any $p \in \partial M = \overline{M} \setminus M$, there exist holomorphic functions $f_1, \ldots, f_n$ on $\overline{M}$ such that $(f_1, \ldots, f_n)$ serve as local coordinates near p. Moreover, by amalgamating a sufficiently small neighbourhood of the image of p to the complex manifold $\overline{M}$, they can consider p as an interior point of a larger finite complete strongly pseudoconvex manifold. Hormander's techniques for Stein manifolds can be applied to $\overline{M}$ to show that for any compact $K \subset M$, the algebra of holomorphic functions on K can be uniformly approximated by functions which are holomorphic on $\overline{M}$. In particular, although a priori, it is only known (by the Newlander-Nirenberg theorem on the existence of local coordinates for integrable complex structures) that local holomorphic coordinates exist, the uniform estimates show that the local coordinates can be assumed to analytically extend to $\overline{M}$ (although not necessarily as global coordinates). Further, since the Steinness of M implies that for any compact subset $K \subset M$, bounded holomorphic functions separate points of K, if follows that global holomorphic functions of $\overline{M}$ separate interior points of $\overline{M}$. By the above comment (that for any bundary point, there exists a larger finite complete strongly pseudoconvex manifold for which it is an interior point), the global holomorphic functions on $\overline{M}$ separate points on $\overline{M}$. The result then follows by the compactness of $\overline{M}$.

The result of Kohn and Rossi-Taylor were used in [8] to provide an intrinsic function theoretic characterization of those open complex manifolds which are biholomorphically equivalent to a smooth bounded strongly pseudoconvex complex submanifold of C^N. Before stating the result, we make the following definition.

(3.3) Definition.

Let ϕ be a smooth bounded strongly pseudoconvex exhaustion of an open complex manifold M. Then ϕ is said to be *smooth and strongly pseudoconvex at infinity* if there exist constants C_k *such that:*

(i) $|\log|d\phi|| \leq C_0$ outside a compact subset

(ii) $\| \nabla^k \phi \| \leq C_k$

(iii) $\| \nabla^k R^\phi \| \leq C_k$

where all norms are taken with respect to the finite Kähler metric $\phi_{i\bar{j}} = \partial^2\phi/\partial z_i \partial z_{\bar{j}}$. $\nabla^k\phi$ and $\nabla^k R^\phi$ refer to the kth covariant derivatives of ϕ and R^ϕ respectively, and R^ϕ is the curvature tensor of the metric $\phi_{i\bar{j}}$. That is, ϕ is said to be smooth and strongly pseudoconvex at infinity if the metric $\phi_{i\bar{j}}$ has bounded geometry and ϕ is smooth relative to this metric.

We are now ready to state the theorem established in [8]:

(3.4) Theorem. Let M be an open complex manifold. Then M is biholomorphically equivalent to a smooth bounded strongly pseudoconvex submanifold of $\mathbf{C}^N \Leftrightarrow$ M admits a smooth bounded strongly pseudoconvex exhaustion ϕ which is smooth and strongly pseudoconvex at infinity.

The idea used in establishing the above theorem is to use the function ϕ to attach a boundary. Since $d\phi \neq 0$ outside a compact, M has a natural collar structure. Define T to be the vector field which is normal to the level sets of ϕ (relative to the metric $\phi_{i\bar{j}}$) and normalized such that $T(\phi) \equiv 1$. The collar structure can be defined by following the integral curves to the vector field T. This makes the complement of a compact into a product manifold, and defines a natural smooth structure on the boundary. The bounds on the curvature and the covariant derivatives of ϕ can be used to show that the metric $\phi_{i\bar{j}}$ extends smoothly to a nondegenerate metric on the boundary, and the complex structure extends smoothly to a strongly pseudoconvex CR structure on the boundary. The resulting compactification is a complete finite strongly pseudoconvex manifold with a smooth Kähler metric $\phi_{i\bar{j}}$. The imbedding results of Rossi and Taylor apply to complete the proof.

§ 4: Weakly Pseudoconvex Domains

The state of knowledge for weakly pseudoconvex domains is much less advanced than for smooth bounded strongly pseudoconvex domains, and it has been a well recognized fact that the weakly pseudoconvex case is in general much more difficult to understand. From a philosophical viewpoint, this can be understood in the fact that the weakly pseudoconvex case is the borderline between the good case (smooth and strongly pseudoconvex) and the bad case (not pseudoconvex), and the closer that one is to the strongly pseudoconvex case (for example, domains of finite type), the easier the analysis becomes. A mathematical understanding of the problem in relation to the $\bar{\partial}$ problem is that this is the point at which uniform positivity of the Levi form is lost, and the subelliptic estimates at the boundary become weaker. On a still more rudimentary level, one no longer has a smooth defining function ϕ for the domain for which $\phi_{i\bar{j}}$ is strictly positive. Recall, for instance, that this was the starting point for

Klembeck in determining the asymptotic behaviour of the curvature for the complete Kähler metric $(-\log(-\phi))_{i\bar{j}}$; similarly, Fefferman's algebraic manipulation to obtain an asymptotic Einstein Kähler metric relied on the fact that $\phi_{i\bar{j}}$ was uniformly invertible. An offshoot of this is that weakly pseudoconvex domains do not possess complete Kähler metrics with the asymptotic growth properties on the curvature tensor satisfied by Klembeck's metric; even the metric, in general, satisfies different asymptotic growth estimates. On the other hand, much of the above program can be carried out, and while the results are still not complete, the outlook is promising.

The weakly pseudoconvex domains which are most completely understood are those of finite type. Many different definitions have been suggested for the concept of finite type; all such definitions attempt to express the degree of degeneracy of the Levi form on the boundary, or the maximum order of contact of analytic subvarieties with the boundary - that is, in some respect they refer to the degree of "flatness" of the boundary. For a survey of the various definitions, their interrelationships and their relationship to the $\bar{\partial}$-Neumann problem, see D'Angelo's article in these proceedings [19]. We will mention here only the definition due to D'Angelo: a point $p \in \partial D$ is said to be of *finite type* if the order of contact of all complex analytic subvarieties with ∂D at p is bounded.

Catlin [14] introduced a more general notion for weakly pseudoconvex domains than that of finite type. He said that a pseudoconvex domain $D \subset\subset \mathbf{C}^n$ satisfied condition P if for every $c > 0$, there exists a smooth, pseudoconvex function ϕ on D such that $0 \leq \phi \leq 1$ on $\bar{D}$ and $[\phi_{i\bar{j}}] \geq C\,[\delta_{ij}]$ on ∂D. In [14], Catlin showed that if $D \subset\subset \mathbf{C}^n$ is weakly pseudoconvex of finite type (as in the sense of D'Angelo), then D satisfies condition P.

With these preliminaries out of the way, we will now summarize what is known for weakly pseudoconvex domains. As in the strongly pseudoconvex case, we begin with the biholomorphic invariance of the boundary. Fefferman's original result on the smooth extension to the boundaries of biholomorphic maps between strongly pseudoconvex domains has been greatly simplified and generalized using a variety of techniques (see e.g. [1, 2, 3, 4, 20, 41]). The most general result relies on an understanding of the Bergman projection operator **P**: that is, the operator which projects the space of L^2 functions on D orthogonally onto the space of L^2 holomorphic functions on D, where the L^2 norms are with respect to the extrinsic Euclidean metric. A domain D is said to satisfy condition R if $\mathbf{P} : C^\infty(\overline{D}) \to C^\infty(\overline{D})$. (By studying the formal setup to the $\bar\partial$-Neumann problem, it is easy to see that a domain D satisfies condition R if the $\bar\partial$ Neumann problem is globally regular on (0, 1) forms; that is, if the canonical solution u to $\bar\partial u = \alpha$ is smooth on $\overline{D}$ whenever α is a smooth (0, 1) form on $\overline{D}$). Bell showed that a proper map $F : D_1 \to D_2$ between two pseudoconvex domains extends smoothly to a map $F : \overline{D}_1 \to \overline{D}_2$ if D_1 satisfies condition R [2]. Further, by an earlier result of Fornaess [24], if F is also a biholomorphism on the interior, then F extends to a diffeomorphism on the boundary. It will follow from the results of Catlin (described below) that if D satisfies condition P, then D satisfies condition R.

From a geometric viewpoint, complete Kähler metrics on weakly pseudoconvex domains will not admit the same nice asymptotic analysis permitted by Klembeck's metrics on the smooth bounded strongly pseudoconvex domain. Indeed, since Klembeck's metrics possessed such stringent asymptotic growth rates, it was possible to regularize the exponential map and to use Jacobi fields to deduce that the regularized exponential map yielded smooth boundary coordinates. In the general case, the asymptotic growth rates can be quite arbitrary, (see e.g. [7]) and the Jacobi equations will not provide the necessary estimates. On the positive side, Mok and Yau [43] established the following result: If D is a bounded pseudoconvex domain in $\mathbf{C}^n$, then D admits a unique complete Einstein Kähler metric. (It follows from Yau's studies on harmonic function theory on complete Riemannian manifolds [49] that the Ricci curvature must be negative). Their proof involves exhausting the pseudoconvex domain by smooth strongly pseudoconvex domains, taking the limit of the Einstein Kähler metrics on the intermediate smooth domains, and obtaining an estimate to ensure that the limiting metric does not degenerate and remains complete. This proof outline is actually illustrative of a general principle for dealing with the remaining cases: first, solve the problem in the good case to obtain an canonically defined solution, and then obtain uniform estimates to ensure that the solution does not degenerate in the limit.

Recently, there has been much progress made in understanding the function theory on weakly pseudoconvex domains. Before describing the main results, we take note of two earlier results due to Kohn and Nirenberg. In [40], they showed that if the quadratic form $Q(\phi,\phi) = \|\phi\|^2 + \|\bar\partial \phi\|^2 + \|\bar\partial^* \phi\|^2$ is compact on (p, q) forms, then the $\bar\partial$ Neumann problem is globally regular on (p, q) forms. In the same paper, they also showed that if $\bar\partial$ Neumann problem satisfies a subelliptic estimate on (p, q) forms

($\|\phi\|_\varepsilon^2 \le C\, Q(\phi, \phi)$ for some constant c independent of ϕ, where $\| \ \|_\varepsilon^2$ is a Sobolev ε-norm, $\varepsilon > 0$), then local regularity holds for the $\bar\partial$ Neumann problem for (p, q) forms; that is, the canonical solution u to $\bar\partial u = \alpha$ is smooth on $\bar{D}$ wherever α is smooth.

Catlin [12, 13, 15] studied the relationship between conditions on the boundary of a weakly pseudoconvex domain D and regularity for the $\bar\partial$-Neumann problem. He showed that a subelliptic estimate holds for the $\bar\partial$-Neumann problem on (p, q) forms if and only if the domain D is of finite q type (D is of finite q type if the order of contact of the boundary with all q dimensional analytic subvarieties is bounded). It follows from the work of Kohn and Nirenberg that the $\bar\partial$-Neumann problem on (p, q) forms is locally regular for domains of finite q type. In [14], Catlin also obtained some results for pseudoconvex domains satisfying condition P. He showed that if D satisfies condition P, then the form Q satisfies a compactness estimate on (0,1) forms; again, it follows from the work of Kohn and Nirenberg, that for smooth pseudoconvex domains satisfying condition P, the $\bar\partial$-Neumann problem is globally regular for (0, 1) forms.

Kohn [37] studied the $\bar\partial$-problem on general smooth bounded pseudoconvex domains using weighted Hilbert spaces. He was able to show that if $D \subset\subset C^n$ is a smooth pseudoconvex domain and if α is a $\bar\partial$ closed (p, q) form which is smooth in $\bar{D}(q \ge 1)$, then for every $k > 0$, there exists a (p, q-1) form u which is in $C^k(\bar{D})$ and satisfies $\bar\partial u = \alpha$. In addition, for every fixed integer k, the solutions u_k satisfy uniform Sobolev estimates $\|u\|_s \le C\,\|\alpha\|_s$ $(s \le k)$ with constants independent of α. Finally, in [38], Kohn showed that by using a diagonalization process suggested by Hörmander, one could find a solution $u \in C^\infty(\bar{D})$ to $\bar\partial u = \alpha$.

Bibliography

1. Bedford, E., Bell, S. and Catlin, D., "*Boundary behavior of proper holomorphic mappings,*" Mich. Math. J. 30 (1983), 107-111.

2. Bell, S., "*Biholomorphic mappings and the* $\bar\partial$-problem," Ann. of Math. (2) 114 (1981), 103-133.

3. Bell, S. and Catlin D., "*Boundary regularity of proper holomorphic mappings,*" Duke Math. J. 49 (1982), 385-396.

4. Bell, S. and Ligocka, E., "*A simplification and extension of Fefferman's theorem on biholomorphic mappings,*" Invent. Math. 57 (1980), 283-289.

5. Bishop, E., "*Mappings of partially analytic spaces,*" Amer. J. Math. 83 (1961), 209-242.

6. Bland, J., "*On the existence of bounded holomorphic functions on complete Kähler manifolds,*" Invent. Math. 81 (1985), 555-566.

7. Bland, J., "*The Einstein-Kähler metric on* $\{|z|^2 + |w|^{2p} < 1\}$," Mich. Math. J. 33 (1986), 209-220.

8. Bland, J., "*Bounded imbeddings of open Kähler manifolds in* $\mathbb{C}^N$," preprint.

9. Bloom, T., and Graham, I., "*A geometric characterization of points of type m on real hypersurfaces,*" J. Diff. Geom. 12 (1977), 171-182.

10. Boutet de Monvel, L., "*Integration des equations de Cauchy-Riemann induites formelles,*" Seminaire Goulaouic-Schwartz, 1974-75.

11. Burns, D., Diederich, K., and Schnider, S., "*Distinguished curves in pseudoconvex boundaries,*" Duke Math. J. 44 (1977), 407-431.

12. Catlin, D., "*Necessary conditions for subellipticity of the* $\bar\partial$*-Neumann problem,*" Ann. of Math. 117 (1983), 147-171.

13. ________, "*Boundary invariants of pseudoconvex domains,*" Ann. of Math. 120 (1984), 529-586.

14. ________, "*Global regularity of the* $\bar\partial$*-Neumann problem,*" Proc. Sym. Pure Math., vol. 41, Amer. Math. Soc., Providence, R.I. (1984), 39-49.

15. ________, "*Subelliptic estimates for the* $\bar\partial$*-Neumann problem on pseudoconvex domains,*" preprint.

16. Cheng, S.Y., and Yau, S.-T., "*On the existence of a complete Kähler metric on noncompact complex manifolds and the regularity of Fefferman's equations,*" Comm. Pure Appl. Math. 33 (1980), 507-544.

17. Chern, S.S. and Moser, J.K., "*Real hypersurfaces in complex manifolds,*" Acta Math. 133 (1974), 219-271.

18. D'Angelo, J., "*Real hypersurfaces, orders of contact, and applications,*" Ann. of Math. (2) 115 (1982), 615-637.

19. ________, "*Finite type conditions for real hypersurfaces in* $\mathbb{C}^n$," preprint.

20. Diederich, K., and Fornaess, J.E., "*Boundary regularity of proper holomorphic mappings,*" Invent. Math. 67 (1982), 363-384.

21. Fefferman, C., "*The Bergman kernel and biholomorphic mappings of pseudoconvex domains,*" Invent. Math. 26 (1974), 1-65.

22. Fefferman, C., "*Monge-Ampère equations, the Bergman kernel, and geometry of pseudoconvex domains,*" Ann. of Math. 103 (1976), 395-416.

23. Folland, G.B. and Kohn, J.J., "*The Neumann problem for the Cauchy-Riemann complex,*" Ann. of Math. Studies, no. 75, Princeton Univ. Press, Princeton, N.J., 1972.

24. Fornaess, J.E., "*Biholomorphic mappings between weakly pseudoconvex domains,*" Pacific J. Math. 74 (1978), 63-65.

25. Grauert, H., "*On Levi's problem and the imbedding of real analytic manifolds,*" Ann. of Math. 68 (1958), 460-472.

26. Greene, R.E. and Wu, H., "*Curvature and complex analysis. I, II, III,*" Bull. Amer. Math. Soc. 77 (1971) 1045-1049; ibid. 78 (1972), 866-870; 79 (1973), 606-608.

27. _________, "*On the subharmonicity and plurisubharmonicity of geodesically convex functions,*" Indiana Univ. Math. J. 22 (1973), 641-653.

28. _________, "*Some function-theoretic properties of noncompact Kähler manifolds,*" Proc. Sym. Pure Math., vol. 27, part II, Amer. Math. Soc., Providence, R.I. (1975), 33-41.

29. _________, "C^∞ *convex functions and manifolds of positive curvature,*" Acta Math. 137 (1976), 209-245.

30. _________, "*Analysis on noncompact Kähler manifolds,*" Proc. Sym. Pure Math., vol. 30, part II, Providence, R.I. (1977), 69-100.

31. _________, "*Gap theorems for noncompact Riemannian manifolds,*" Duke Math. J. 49 (1982), 731-756.

32. Hörmander, L., "L^2 estimates and existence theorems for the $\bar\partial$ operator," Acta Math. 113 (1965), 89-152.

33. Hörmander, L., "*An introduction to complex analysis in several variables,*" Van Nostrand, Princeton, N.J. 1966.

34. Klembeck, P., "*Kähler metrics of negative curvature, the Bergman metric near the boundary and the Kobayashi metric on smooth bounded strictly pseudoconvex sets,*" Indiana Math. J. 27 (2) (1978), 275-282.

35. Kohn, J.J., "*Harmonic integrals on strongly pseudoconvex manifolds, I, II*", Ann. of Math. (2) 78 (1963), 112-148; ibid. (2) 79 (1964), 450-472.

36. ________, "*Boundary behaviour of $\bar{\partial}$ on weakly pseudoconvex manifolds of dimension two,*" J. Diff. Geom. 6 (1972), 523-542.

37. ________, "*Global regularity for $\bar{\partial}$ on weakly pseudoconvex manifolds,*" Trans. Amer. Math. Soc. 181 (1973), 273-292.

38. ________, "*Methods of partial differential equations,*" Proc. Sym. Pure Math., vol. 30, Part I, Amer. Math. Soc., Providence, R.I. (1977), 215-237.

39. ________, "*Subellipticity of the* $\bar{\partial}$-Neumann problem on pseudoconvex domains: Sufficient conditions," Acta Math. 142 (1979), 79-122.

40. Kohn, J.J. and Nirenberg, L., "*Noncoercive boundary value problems,*" Comm. Pure Appl. Math. 18 (1965), 443-492.

41. Lempert, L., "*La métrique de Kobayashi et la répresentation des domaine sur la boule,*" Bull. Soc. Math. France 109 (1981), 427-474.

42 Mok, N., Siu, Y.T. and Yau, S.-T., "*The Poincaré-Lelong equation on complete Kähler manifolds,*" Compositio Math. 44 (1981), no. 1-3, 183-218.

43. Mok, N. and Yau, S.-T., "*Completeness of the Kähler Einstein metric on bounded domains and the characterization of domains of holomorphy by curvature conditions,*" Proc. Sym. Pure Math., Vol. 39, Part I, Providence, R.I. (1983), 41-59.

44. Narasimhan, R., "*Imbedding of holomorphically complete complex spaces,*" Amer. J. Math. 82 (1960), 917-934.

45. Remmert, R., "*Sur les espaces analytiques holomorphiquement séparables et holomorphiquement convexes,*" C.R. Acad. Sci. Paris 243 (1956), 118-121.

46. Rossi, H. and Taylor J., "*On algebras of holomorphic functions on finite pseudoconvex manifolds,*" J. Functional Anal. 24 (1977), 11-31.

47. Siu, Y.T., and Yau, S.-T., "*Complete Kähler manifolds with nonpositive curvature of faster than quadratic decay,*" Ann. of Math. 105 (1977), 225-264.

48. Webster, S., "*Kähler metrics associated to a real hypersurface,*" Comment. Math. Helvetici 52 (1977), 235-250.

49. Yau, S.-T., "*Harmonic Functions on complete Riemannian manifolds,*" Comm. Pure Appl. Math. 28 (1975), 201-228.

University of Toronto
Toronto, Ontario, Canada
M5S 1A1

A CHARACTERIZATION OF CP^n BY ITS AUTOMORPHISM GROUP

by

J. Bland, *University of Toronto*
T. Duchamp, *University of Washington*
M. Kalka, *Tulane University*

§ 0. Introduction.

In this note we extend the results of Greene and Krantz [**GK**] to the case of compact manifolds. To be more precise, let M be an n-dimensional complex manifold, let p be a point of M and let G be a group of biholomorphisms of M fixing p. By differentiation the group G acts linearly on the complex tangent space M_p and, therefore, on the complex projective space $\boldsymbol{CP}(M_p)$ of complex directions at p. If the action of G on $\boldsymbol{CP}(M_p)$ is transitive then G is said to *act transitively on complex directions.* Our main result is the following theorem, which answers affirmatively a conjecture of Krantz:

0.1. Theorem. *Let M be a compact complex manifold and suppose that there is a compact group of biholomorphisms of M fixing a point p and acting transitively on complex directions at p. Then M is biholomorphic to $\boldsymbol{CP}^n$.*

0.2 Remarks. i) In the case where M is not compact Greene and Krantz show that M is either biholomorphic to the unit ball in $\boldsymbol{C}^n$ or to $\boldsymbol{C}^n$ itself. Although not in the statement of their results, they show that the biholomorphism can be made to send p to the origin and to identify G with a linear subgroup of the unitary group U(n) with its usual linear action on $\boldsymbol{C}^n$ and that this biholomorphism is unique up to composition with a unitary transformation.

ii) In [**GK**] the apparently stronger assumption, that G acts transitively on *real* directions at p, is made. However, this is unnecessary by the following argument: Since G is compact, we can arrange for G to act isometrically with respect to a Hermitian metric on M. Now let S_p be the unit tangent sphere to M at p. Then G acts on S_p and the Hopf map $S_p \longrightarrow \boldsymbol{CP}(M_p)$ is G-equivariant. Assume that G acts transitively on complex directions; then its action on the unit sphere is transverse to the fibers on the Hopf map. Either G acts transitively on S_p (and the result follows) or the orbits of G cover $\boldsymbol{CP}(M_p)$. However, the second case is impossible since $\boldsymbol{CP}(M_p)$ is simply connected so such a cover would give a section of the Hopf map; but it is well known that none exist.

iii) Since the space $\boldsymbol{CP}(M_p)$ is connected, the connected component of the identity in G will act transitively on complex directions whenever G does. Therefore, without loss of generality we may assume that G is connected.

iv) After this work was completed we learned that the above theorem is a weaker version of a theorem of E. Oeljeklaus [**O**]:

Let M be a compact complex manifold and G' any group of biholomorphism on M. If there is a point $p \in M$ fixed by G' and a neighborhood U_p in M such that $U_p \setminus \{p\}$ is an orbit of G' then M is biholomorphic to CP^n.

Oeljeklaus has also informed us that he and A. Huckleberry have obtained classification theorems for complex manifolds equipped with group actions by biholomorphisms (see [HO] and the references therein).

Acknowledgement. We wish to thank S. Krantz for posing the problem of characterizing CP^n by its group of biholomorphisms to us, J. Morrow who explained his work on compactifications of C^2 to us and E. Oeljeklaus for informing us of his work and the work of Huckleberry.

§ 1. Outline of the Proof.

In this section the proof of the main theorem is outlined. For the remainder of the paper M will be an n-dimensional compact complex manifold with G a compact group of biholomorphisms fixing a point p and acting transitively on complex directions at p. We will give M a Hermitian metric with respect to which G acts by isometries.

Let C_p denote the cut locus of p (see [K] for elementary facts concerning cut loci). The results in [BM] show that if M is obtained from C^n by attaching a copy of CP^{n-1} then the resulting manifold is biholomorphic to CP^n. Hence, it suffices to show that $M \setminus C_p$ is biholomorphic to C^n and that C_p is a complex hypersurface which is biholomorphic to CP^{n-1}.

The first step in the proof, presented in section 2, is to show that C_p is a compact, connected, complex hypersurface whose complement is biholomorphic to C^n.

The next step proceeds as follows. The biholomorphism between $M \setminus C_p$ and C^n allows us to define a singular holomorphic foliation of $M \setminus C_p$ (just take the complex lines in C^n passing through the origin). The leaves of this foliation are holomorphically parameterized by their (one dimensional) tangent spaces at the point p, i.e. by points in $CP(M_p)$. We show that the closure of each leaf is a smooth complex curve, biholomorphic to CP^1 which intersects the cut locus transversely at a single point. The closures of disjoint lines will be shown to be disjoint. This foliation furnishes the biholomorphism between $CP(M_p)$ and C_p needed to apply the results of [BM] and completing the proof of the theorem.

§ 2. The cut locus of p.

We begin the analysis of C_p with some elementary observations. By Remark 0.2.ii, G acts transitively on the unit tangent sphere at p. It follows that G acts transitively on the set of geodesics starting at p. Since each geodesic from p intersects C_p, it follows that G acts transitively on C_p and that the distance from p to any point in C_p is a constant which we will choose to be 1. In particular, the cut locus, being a G-orbit, is a smooth, connected submanifold of M. Let $\exp_p$ be the exponential map at p, B_p the unit ball in M_p and S_p the unit sphere in M_p. Then $\exp_p$ restricts to a map from S_p to C_p which, by transitivity again, has constant rank. The following lemma summarizes several important properties of the exponential map and is essentially contained in [K], p 100.

2.1 Lemma. *The fibers of the map $\exp_p : S_p \longrightarrow C_p$ are in 1-1 correspondence with the geodesics joining p to a fixed point in the image of $\exp_p$. The manifold M is obtained by attaching the ball B_p to C_p via the above fibration.*

2.2 Proposition. *The cut locus of p, C_p, is a complex hypersurface of M whose complement is biholomorphic to $\boldsymbol{C}^n$.*

Proof. By [**GK**] the complement of the cut locus is biholomorphic to the unit ball or to complex n-space and, therefore, supports non-constant holomorphic functions. Since C_p is a real submanifold it has pure Hausdorff dimension. Therefore, by the result of [**S**],Lemma 3(i), if C_p had real codimension greater than 2 then M would support non-constant holomorphic functions. However, M is compact; hence C_p has real codimension either 1 or 2.

Suppose that the real codimension of the cut locus is 1. Then the map $S_p \longrightarrow C_p$ induced by the exponential map is a finite covering map. But the number of sheets is precisely the number of geodesics starting at p and terminating at a point q in C_p. The geodesics joining p to the cut locus are extremals of the arc length functional with variable end point on C_p. As such they satisfy the transversality conditions of the calculus of variations, i.e. they intersect C_p at right angles. Since C_p is a real hypersurface the covering must be 2 to 1. Now a fixed point free action of $\boldsymbol{Z}_2$ on an odd dimensional sphere is orientation preserving (since the $\boldsymbol{Z}_2$ action has no fixed points it follows from the Lefschetz fixed point theorem that the action of $\boldsymbol{Z}_2$ on the top dimensional homology group is trivial), hence, the cut locus is orientable. On the other hand the normal bundle of the cut locus is one sided by the following argument. By transitivity of the action of the connected group G on the sphere, S_p, any two geodesics starting at p can be connected by a family of geodesics starting at p, showing that C_p is one-sided. Because there are no one-sided, orientable hypersurfaces in an orientable manifold the codimension of C_p cannot be 1 and is necessarily 2.

By [**GK**] the complement of the cut locus is biholomorphic to complex Euclidean space or to the ball. Since the ball supports bounded, non-constant holomorphic functions and by [**S**],Lemma 3(ii) such functions would extend to all of M contradicting the compactness of M, the complement of C_p is biholomorphic to $\boldsymbol{C}^n$.

It remains to show that C_p is a complex hypersurface. To see this let $G^{\mathbf{C}}$ denote the complexification of G. Then $G^{\mathbf{C}}$ acts holomorphically on M and the orbits of $G^{\mathbf{C}}$ are complex submanifolds of M. Since the complement of the cut locus can be identified with complex n-space in such a way that G is a subgroup of U(n) it follows that $G^{\mathbf{C}}$ leaves the complement of the cut locus invariant. This in turn shows that $G^{\mathbf{C}}$ leaves C_p invariant. The group G acts transitively on C_p therefore C_p is on orbit of $G^{\mathbf{C}}$ and a complex submanifold of M. **QED**

§ 3. A foliation of M by $\boldsymbol{CP}^1$'s.

In this section we present the proof of Theorem 0.1. The idea is to construct a foliation, $\mathcal{F}$, of M by $\boldsymbol{CP}^1$'s which intersect C_p transversely and to use the foliation to construct a biholomorphism between C_p and $\boldsymbol{CP}^{n-1}$. Theorem 0.1 then follows from the results of [**BM**].

To construct the foliation $\mathcal{F}$ start with the observation that since $M \setminus C_p$ is biholomorphic to $\boldsymbol{C}^n$, it is naturally foliated by lines (identify p with the origin and consider the foliation of C^n by lines through the origin). This foliation is singular at p, but becomes smooth after blowing up p. We will show that the closure in M of each line is smooth and intersects C_p transversely in a single point. The foliation $\mathcal{F}$ is then the foliation whose leaves are the closures of the above lines. That $\mathcal{F}$ is holomorphic is clear because it is holomorphic on the open dense set, $M \setminus C_p$. To construct a biholomorphism between the complex projective space $\mathrm{CP}(M_p)$ and C_p just map the complex line $[\mathbf{v}] \subset M_p$, $\mathbf{v} \in M_p$ to the point of intersection of C_p and the leaf of $\mathcal{F}$ whose tangent space at p is $[\mathbf{v}]$

3.1 Remark. The biholomorphism between C^n and $M \setminus C_p$ can be extended to a biholomorphism between CP^n and M without using the results of [BM] as follows. Let q be any point in the hyperplane at infinity, let $[\mathbf{v}_q] \subset C^n$ be the unique line whose closure in CP^n contains q and let L_q be the leaf of $\mathcal{F}$ whose tangent space at p is $[\mathbf{v}]$ (recall we are identifying M_p and C^n). The image of Q is defined to be the unique point in the intersection $L_q \cap C_p$.

The proof that the closures of the above mentioned complex lines intersect the cut locus transversely and in a single point requires the introduction of an intermediate foliation. To construct a leaf consider any non-zero vector $\mathbf{v}$ in the tangent space M_p. (For convenience we will identify $M \setminus C_p$ with C^n and M_p with C^n. This abuse of notation will cause no confusion and simplifies the notation.)

Let $K_\mathbf{v}$ be the subgroup of G which leaves $\mathbf{v}$ fixed and $V_\mathbf{v}$ the linear subspace consisting of all vectors fixed by $K_\mathbf{v}$. Note that if $\mathbf{w}$ is in $V_\mathbf{v}$ than the geodesic $t \to \exp_p(t\mathbf{w})$ is left fixed by $K_\mathbf{v}$. (Because G acts linearly on $V_\mathbf{v}$ we may identify the tangent space of $V_\mathbf{v}$ at the origin with $V_\mathbf{v}$ itself.) Therefore, the vector space $V_\mathbf{v}$ is a union of the geodesic segments $t \to \exp_p(t\mathbf{w})$, $0 \le t \le 1$, $\mathbf{w} \in V_\mathbf{v}$ and the closure of $V_\mathbf{v}$, which we will denote by the symbol $N_\mathbf{v}$, is the union of the geodesics $t \to \exp_p(t\mathbf{w})$, $0 \le t$, $\mathbf{w} \in V_\mathbf{v}$.

We claim that the complex dimension of $V_\mathbf{v}$ is either 1 or 2. Clearly, because $V_\mathbf{v}$ contains the line generated by $\mathbf{v}$, its dimension is at least 1.

To see that the dimension of $V_\mathbf{v}$ cannot exceed 2 proceed as follows. It is easy to see that for any $h \in G$, the conditions $h\mathbf{v} = \mathbf{v}$ and $ghg^{-1}g\mathbf{v} = g\mathbf{v}$ are equivalent. The next two equalities follow immediately:

$$K_{g\mathbf{v}} = gK_\mathbf{v}g^{-1}$$

and

(3.2) $$V_{g\mathbf{v}} = g(V_\mathbf{v})$$

If $H_\mathbf{v}$ is the subgroup of G which leaves the space $V_\mathbf{v}$ invariant then above equalities can be used to prove the chain of equivalences

$$g \in H_\mathbf{v} \iff g(V_\mathbf{v}) = V_\mathbf{v} \iff V_{g\mathbf{v}} = V_\mathbf{v} \iff K_{g\mathbf{v}} = K_\mathbf{v} \iff gK_\mathbf{v}g^{-1} = K_\mathbf{v}.$$

The last equivalence shows that $K_\mathbf{v}$ is a normal subgroup of $H_\mathbf{v}$. Further, for any automorphism $g \in G$ such that $g^{-1}\mathbf{v} \in V_\mathbf{v}$ the condition $g^{-1} \in H_\mathbf{v}$ is easily seen to hold. This gives the following characterization of $H_\mathbf{v}$:

(3.3) $$H_{\mathbf{v}} = \{g \in G | g^{-1}\mathbf{v} \in V_{\mathbf{v}}\}$$

From (3.3) and the assumption that G acts transitively on directions it follows that $H_{\mathbf{v}}$ acts transitively on directions in $V_{\mathbf{v}}$ and that the group $H_{\mathbf{v}}/K_{\mathbf{v}}$ acts transitively and freely on the unit sphere in $V_{\mathbf{v}}$. But then the group $H_{\mathbf{v}}/K_{\mathbf{v}}$ is diffeomorphic to a sphere and therefore either the one or the three sphere (since by linearity of the G action on $\boldsymbol{C}^n$, $V_{\mathbf{v}}$ contains the complex line generated by $\mathbf{v}$, $H_{\mathbf{v}}/K_{\mathbf{v}}$ is not the zero sphere) and the complex dimension of $V_{\mathbf{v}}$ is either 1 or 2.

Observe that since $K_{\mathbf{v}}$ is normal in $H_{\mathbf{v}}$ formula (3.2) shows that given two vectors $\mathbf{v}$ and $\mathbf{w}$ the spaces $V_{\mathbf{v}}$ and $V_{\mathbf{w}}$ are either disjoint or equal.

We next show that the set $N_{\mathbf{v}}$ is a smooth, compact, complex submanifold of M which intersects C_p transversely and that the family, $N_{\mathbf{v}}$, $\mathbf{v} \in M_p$ forms a smooth foliation of M which is singular only at p. Since $V_{\mathbf{v}}$ is a complex analytic submanifold, its closure will automatically be complex analytic if only we show that is smooth.

First note that since the Lie group $H_{\mathbf{v}}$, and hence the connected component of the identity in $H_{\mathbf{v}}$, acts transitively on the unit tangent vectors to $N_{\mathbf{v}}$ at p, it acts transitively on the intersection $N_{\mathbf{v}} \cap C_p$. Therefore, N_p intersects C_p in an orbit of a connected Lie group, hence, the intersection, $N_{\mathbf{v}} \cap C_p$, is a connected submanifold. To show that $N_{\mathbf{v}}$ is a submanifold it is sufficient to show that it is the union of all geodesics which intersect $N_{\mathbf{v}} \cap C_p$ and are normal to C_p. To see this let q be a point on $N_{\mathbf{v}} \cap C_p$, let $t \rightarrow \exp_p(t\mathbf{w})$, $\mathbf{w} \in V_{\mathbf{v}}$ be a geodesic containing q and let $\mathbf{w}_q$ be the tangent vector to the geodesic at q. Since every element of $K_{\mathbf{v}}$ fixes the geodesic (and therefore $\mathbf{w}_q$) it follows that $K_{\mathbf{v}}$ fixes all normal vectors to C_p at q fixed (this space is one dimensional and has $\mathbf{w}_q$ as basis). But $K_{\mathbf{v}}$ acts by isometries and therefore fixes all geodesics normal to C_p at q. From the definition of $V_{\mathbf{v}}$ it follows that these geodesics are all in $N_{\mathbf{v}}$, as was to be shown.

To see that the family of all forms a foliation it suffices to show that for $\mathbf{v}, \mathbf{w} \in M_p$, the submanifolds and $N_{\mathbf{w}}$ are either equal or intersect in the single point p. If and $N_{\mathbf{w}}$ intersect at a point in $M \setminus C_p$ then they intersect along the unique geodesic joining the point to p and following this geodesic to a point in C_p and therefore intersect C_p. But the analysis of the above paragraph shows that if and $N_{\mathbf{w}}$ intersect in C_p they are equal.

The manifold M is obtained from the complex one or two dimensional vector space, $V_{\mathbf{v}}$, by attaching the smooth, compact, connected complex manifold $N_{\mathbf{v}} \cap C_p$. If the dimension of $V_{\mathbf{v}}$ i8s one then we have already shown that the closure of each line in $M \setminus C_p$ containing p is biholomorphic to $\boldsymbol{CP}^1$ and that the closures of each line intersects C_p transversely in a single point.

If the dimension of $V_{\mathbf{v}}$ is 2 apply the result of Morrow ([M],Lemma 5) to conclude that $N_{\mathbf{v}}$ is biholomorphic to $\boldsymbol{CP}^2$ and identify $N_{\mathbf{v}} \cap C_p$ with the $\boldsymbol{CP}^1$ at infinity in $\boldsymbol{CP}^2$. The closures of complex lines through the point p in $\boldsymbol{CP}^2$ are all smooth $\boldsymbol{CP}^1$'s which intersect the hyperplane at infinity transversely in single points and form a foliation of $\boldsymbol{CP}^2$ which is singular only at p. Because the intermediate foliation intersects the cut locus transversely so does the foliation of M by $\boldsymbol{CP}^1$'s.

§ References.

[BM] L. Brenton and J. Morrow, *Compactifications of C^n*, Trans. AMS, **246**(1978), 139-153.

[GK] R. E. Greene and S. G. Krantz, *Characterization of complex manifolds by the isotropy subgroups of their automorphism groups*, preprint, (1984).

[HO] A.T. Huckleberry and E. Oeljeklaus, *Classification theorems for almost homogeneous spaces*, Institut Elie Cartan **9** n°, January, 1984.

[K] S. Kobayashi, *On conjugate and cut loci*, Studies in global geometry and analysis, S. S. Chern ed., MAA Studies in Mathematics Vol. 4, 1967.96-122

[M] J. Morrow, *Minimal normal compactifications of C^2*, Rice University Studies**59**(1973), 97-112.

[O] E. Oeljeklaus, *Ein Hebbarkeitssatz für Automorphismengruppen kompakter Mannigfaltigkeiten*, Math. Ann. **190**(1970), 154-166.

[S] B. Shiffman, *On the removal of singularities of analytic sets*, Michigan Math. Journal **15**(1968), 111-120.

Proper Mappings Between Balls in C^n

Joseph A. Cima and Ted Suffridge

1. Introduction.

This paper will consider some problems in the theory of proper mappings from the unit ball $\mathbb{B}_n \subseteq C^n$ into $\mathbb{B}_k$ for $k > n \geq 1$. It will review the material which is available in the literature (or in preprint form). It will present some results of ongoing work of the authors. Finally, we discuss some open problems of interest to us. In addition to the list of literature we will include a specific listing of some proper mappings. There are extensions of some techniques to strongly pseudoconvex domains [1,5,11] but we do not address these questions.

2. Notation.

A holomorphic mapping $f : \mathbb{B}_n \to \mathbb{B}_k$ is proper if $f^{-1}(K)$ is compact in $\mathbb{B}_n$ for each K compact in $\mathbb{B}_k$. This is equivalent to the statement that $||f(z_m)||_k \to 1$ whenever $||z_m||_n \to 1$. We shall drop the subscripts on the norms. If there is a C^k function F defined on $\overline{\mathbb{B}_n}$ such that F restricted to $\mathbb{B}_n$ equals f, then we will say that f is C^k on $\overline{\mathbb{B}_n}$ and write $f \in C^k(\overline{\mathbb{B}_n})$. Note that the norm of every proper map f (on $\mathbb{B}_n \to \mathbb{B}_k$) is continuous on $\overline{\mathbb{B}_n}$. For $v \in \mathbb{B}_n$ the mapping

$$T_v(Z) = \frac{1}{1 + \langle Z,v\rangle}\left[(\sqrt{1 - ||v||^2})Z + \left(1 + \frac{\langle Z,v\rangle}{1 + \sqrt{1 - ||v||^2}}\right)v\right]$$

is a holomorphic automorphism of $\mathbb{B}_n$ onto $\mathbb{B}_n$ taking 0 to v. For $f : \mathbb{B}_n \to \mathbb{B}_k$ proper we will say that g is equivalent to f if there exist $\psi \in \mathrm{Aut}(\mathbb{B}_k)$ and $\phi \in \mathrm{Aut}(\mathbb{B}_n)$ such that

$$g(Z) = \psi \circ f \circ \phi(Z).$$

Then g is also a proper mapping and we shall be interested in classifying proper maps using this equivalence. If f is a proper mapping of $\mathbb{B}_n \to \mathbb{B}_k$ and f is $C^\ell(\overline{\mathbb{B}_n})$ then we write $f \in I^\ell(n,k)$.

3. Some examples.

Perhaps the best way to begin is to give a list of some examples. These will be useful to give the reader some feeling for such mappings.

Also, and for our purposes more important, we wish to refer back to these examples. Some of them are key to pointing out similarities and others differences in the behavior of various cases.

(3.1) $f(Z) = (z_1,\dots,z_n,0,\dots,0)$ $\quad \mathbb{B}_n \to \mathbb{B}_k$

(3.2) $f(Z) = (z_1^2,\sqrt{2}z_1z_2,z_2^2)$ $\quad \mathbb{B}_2 \to \mathbb{B}_3$

(3.3) $f(Z) = (z_1,\dots,z_{n-1},z_1z_n,z_2z_n,\dots,z_n^2)$ $\quad \mathbb{B}_n \to \mathbb{B}_{2n-1}$

(3.4) $f(Z) = (w_1,w_2,w_3,\dots)$ $\quad \mathbb{B}_2 \to \ell^2$

In the last example we choose $0 < r < 1$, and then

$$w_1 = \sqrt{r}\, z_1$$

$$w_2 = r\, z_2$$

$$w_j = \frac{r(1-r)\dots(n-1-r)}{\ell!(n-\ell)!} z_1^{\ell} z_2^{n-\ell}$$

$$3 \le j = \frac{n(n+1)}{2}, \ell < \frac{(n+1)(n+2)}{2}$$

and $\ell \le 2$. In general if $\sum_1^{\infty} c_m^2 ||z||^{2m}$ is any power series with non-negative coefficients that sums to one when $||z||$ equals one, then one obtains a proper holomorphic map from $\mathbb{B}_n$ to ℓ^2 by taking square roots of each term in the sum and listing these as the components. Notice that all the mappings listed except (3.4), are holomorphic on $\overline{\mathbb{B}_n}$. In the later sections we will discuss some constructions of proper mappings.

4. The cases $k = n + 1$.

In the first part of this section we will discuss the results of S. Webster and J. Faran. One of the techniques used in their work is due to H. Lewy [10]. There is a generalized form of this result and so rather than present it here we will include it in a later section. We say that H is an affine subspace of $\mathbb{B}_n$ if $H = A \cap \mathbb{B}_n$, where A is an affine subspace of C^n. Theorem (4.1) below is due to S. Webster [12], and Theorem (4.2) is due to J. Faran [3].

Theorem 4.1. *If* $n \ge 3$ *and* $f \in I^3(n,n+1)$ *then* f *maps* $\mathbb{B}_n$ *into an* n *dimensional subspace of* $\mathbb{B}_{n+1}$.

Because automorphisms of $\mathbb{B}_n$ preserve affine spaces (and because of the equidimensional result of H. Alexander) this result implies that such maps are equivalent to ones of the form (3.1). It is obvious from the form of the automorphisms that example (3.2) is not equivalent to one of type (3.1).

Theorem 4.2. *If* $f \in I^3(2,3)$ *then* f *is equivalent to one of the maps* (3.1), (3.2), (z_1, z_1z_2, z_2^2) *or* $(z_1^3, \sqrt{3}\, z_1z_2, z_2^3)$.

This result was very surprising. At the conference (of these proceedings) John D'Angelo has provided some useful insight into understanding the difference between the $n = 2$ and $n \geq 3$ cases above.

In [2] we have proved a key lemma and extended the Lewy Theorem to a more general setting. From these results we improve Theorem 4.1.

Lemma 4.3. *Let* $f \in I^2(n,n+1)$ *and* $u,v \in \partial \mathbb{B}_n = S$ *with* $\langle u,v\rangle = 0$. *Then*

(a) $\langle D^j f(u)(v^j), f(u)\rangle = 0 \qquad j = 1,2$

and

(b) $||Df(u)(v)||^2 = \langle Df(u)(u), f(u)\rangle \geq 1$.

If f *is analytic on* $\overline{\mathbb{B}}_n$ *the equation* (a) *is valid for all* $j \geq 1$.

Proof. Let θ be real and λ a complex number of modulus one. The vector $w = (\cos\theta)u + \lambda(\sin\theta)v$ has length one and

$$f(w) = f(u) + Df(u)(w-u) + \frac{1}{2} D^2 f(u)(w-u)^2 + R(w,u)$$

where $R(w,u)(||w - u||^2)^{-1} \to 0$ as $||w - u|| \to 0$. By equating coefficients of the parameters θ and λ in the equations

$$1 = \langle f(u), f(u)\rangle = \langle f(w), f(w)\rangle$$

we conclude that (a) is valid. Examining the θ^2 terms in this equality one concludes

$$||Df(u)(v)||^2 = \mathrm{Re}\langle f(u), Df(u)(v)\rangle .$$

Part of (b) is obtained now by using

$$0 = \left(\frac{d}{dt}||f(e^{it}u)||^2\right)_{t=0} = 2\,\mathrm{Re}(-i\langle f(u), Df(u)(u)\rangle)$$

By the Schwartz lemma

$$||f(ru)||^2 \leq r^2$$

so

$$(1-r^2) \le \int_r^1 \left|\frac{d}{d\rho}||f(\rho u)||^2\right| d\rho = 2\int_r^1 \text{Re} \quad \langle Df(\rho u)(u), f(\rho u)\rangle \, d\rho \; .$$

This implies $\langle Df(u)(u), f(u)\rangle \ge 1$ and finishes the proof of (b).

The lemma below is our generalization of Lewy's Theorem. It is valid in a more general setting involving strictly pseudoconvex domains [1]. If $u = (u_1,\ldots,u_n) \in S$ then $v^j = (\bar{u}_j, 0, \ldots, -\bar{u}_1, 0, \ldots, 0)$ $j = 2,\ldots,n$ are in S and $\langle u, v^j\rangle = 0$.

Lemma 4.4. *Assume N is an open set in C^n, $N \cap S \neq \emptyset$, $f : \mathbb{B}_n \to \mathbb{B}_k$ with $f \in C^2(\overline{\mathbb{B}_n \cap N})$ and*

$$f : N \cap S \to \partial\mathbb{B}_k$$

where $n + 1 \le k \le \frac{n(n+1)}{2}$. Further, assume there exists a set K of $k-n$ pairs of integers (ℓ,p), $2 \le \ell \le n$, $2 \le p \le n$ such that for $u \in N \cap S$, the set of k vectors

$$\{f(u)\} \cup \{Df(u)(v^j)\}_{j=2}^n \cup \{D^2f(u)(v^\ell, v^p) : (\ell,p) \in K\}$$

is linearly independent. Then there is a mapping F, holomorphic on N with values in C^k that coincides with f on $\mathbb{B}_n \cap N$.

Proof. In this lemma the key step is an application of Cramer's rule. Let $\alpha_0 \in C$, $|\alpha_0| = 1$ and $u \in S \cap N$ with $\alpha_0 u \in N$. Consider for ρ small the set of α with $(|\alpha_0-\alpha| \le \rho) \cap (|\alpha| \ge 1)$ and the k equations

$$\langle w, f(\tfrac{1}{\alpha} u)\rangle = 1$$

$$\langle w, Df(\tfrac{1}{\alpha} u)(v^j)\rangle = 0 \qquad 2 \le j \le n$$

$$\langle w, D^2f(\tfrac{1}{\alpha} u)(v^\ell, v^p)\rangle = 0 \qquad (\ell,p) \in K\,.$$

By Cramer's rule there is a function $w = F(\alpha u)$ which provides a holomorphic continuation of f through the slice determined by $\alpha_0 u$. Thus f has an extension to a larger open set. It is routine to show that the holomorphy of f on slices implies that this extension is holomorphic as a function of several complex variables (i.e. as a function on $N \subset C^n$).

In section 6 we will prove a continuation theorem for rational proper mappings. But prior to proving this result we proved the following theorem [2].

Theorem 4.5. *If $f \in I^2(n,n+1)$ and $n \ge 3$ then f is equivalent* *to a mapping $\frac{p}{q}$, where p is a polynomial mapping of degree $p \leq 3$ and q is a polynomial, of degree $\leq p-1$.*

Note that this improves the theorem of Webster (Theorem 4.1). It does not yet improve the result of Faran. We return to this question in section 6.

Although most of these results are stated for the $\mathbb{B}_n \to \mathbb{B}_{n+1}$ case they are valid with suitable restrictions for proper maps from $\mathbb{B}_n \to \mathbb{B}_k$. We pose the following question. Assume f is a proper map of B_n into B_k and each component of f is a polynomial. For fixed n and k, is there an upper bound on the degree of f? Note that one can produce proper polynomial maps of arbitrarily high degree by letting k increase. Namely, write $1 \equiv ||z||^{2N}$. Then, by expanding $||z||^{2N}$ by the multinomial theorem, and calling each term $|f_j|^2$, we obtain a proper map f to B_k. Here k depends on both n and N. By reducing the expansion modulo the ideal $||z||^2 - 1$, one can obtain proper maps to lower dimensional balls. This is how one obtains example (3.5).

John D'Angelo has conjectured that by combining this process with the application of automorphisms of balls in intermediate dimensions, one obtains all proper holomorphic mappings that extend holomorphically past the boundary.

We end this section with an interesting tool that has been useful in our work.

Theorem 4.6. *Assume f satisfies the conditions of Lemma 4.4 and A is an n-1 dimensional affine subspace of C^n. There exists a k-1 dimensional subspace D of C^k such that*

$$f : A \cap B_n \to D \cap \mathbb{B}_k .$$

Proof. Since automorphisms map affine subspaces into affine subspaces we may assume without loss of generality that $0 \in A$ and $f(0) = 0$. Fix $u \in N$ and consider pairs (v_j, ℓ_j), $j = 1,2,\ldots,k$ where v_j is a unit vector orthogonal to u and ℓ_j is a positive integer. The notation $(v_j)^{\ell_j} = (v_j,\ldots,v_j)$, ℓ_j times is used. The holomorphic function

$$g(\lambda) = \det \begin{bmatrix} D^{\ell_1} f(\lambda u)(v_1)^{\ell_1} \\ \\ D^{\ell_k} f(\lambda u)(v_k)^{\ell_k} \end{bmatrix} \qquad \lambda \in C, \ |\lambda| < 1$$

is zero on an arc of the unit circle. Hence $g(\lambda) \equiv 0$ and

$$D = \operatorname{span}\{D^j f(0)(v)^\ell \; ; \; \ell = 1,2,3,\ldots, \langle u,v\rangle = 0\}$$

has dimension $\leq k-1$. But the power series expansion

$$f(w) = \sum_{j=1}^{\infty} \frac{1}{j!} D^j f(0)(w)^j$$

shows that the range of f on the space $\{v \in \mathbb{B}_n : \langle v,u\rangle = 0\}$ is of dimension $k-1$.

5. Results on discontinuous mappings.

Several papers have been done by Josep Globevnik and L. Stout [6,7]. They study proper maps from $\mathbb{B}_1$ into $\mathbb{B}_n$, $n \geq 2$. In the higher dimensional case E. Low [11] has established the following result.

Theorem 5.1. Let $f : S = \partial\mathbb{B}_n \to \mathbb{B}_{2N}$ be a continuous mapping and $\varepsilon > 0$, $R < 1$. Then there exists a (nonconstant) mapping $h : \overline{\mathbb{B}}_n \to C^{2N}$ which is holomorphic in $\mathbb{B}_n$ such that $|f(z) + h(z)| = 1$ for all $z \in S$ (and $|h(\xi)| \leq \varepsilon$ when $|\xi| < R$).

This is a significant result. The proof is highly computational but there are two ideas involved in the proof that we wish to comment on. The function h is constructed by applying the approximation technique developed by Hakim and Sibony [8]. First, let us comment on the integer N appearing in the range of the mapping h. For $z,\zeta \in S(=\partial\mathbb{B}_n)$ set $\delta(z,\zeta) = |z-\zeta|/\sqrt{2}$ and denote the ball about z_0 of radius r in this metric by $B(z_0,r)$. The following covering result is known. For $\alpha > 1$ there exists an integer $N = N(n,\alpha)$ such that for $r > 0$ there exist N families $F_1,\ldots,F_N$ of balls with radii αr

$$F_i = \{B(z_{i,j};\alpha r);\ j = 1,2,\ldots,N_i\}$$

$$B(z_{i,j};\alpha r) \cap B(z_{1,k};\alpha r) = \emptyset \qquad j \neq k$$

and

$$S = \bigcup_{i,j} B(z_{ij},r) \quad .$$

We do not know of any good estimates on the size of N and it would be very useful to obtain such estimates.

The second tool that is important in this proof is the following. Let $\{E_j;\ j = 1,2,\ldots,2N\}$ be the usual (canonical) basis in C^{2N} and let T_w be the complex tangent space at w. Then $T_w \cap \mathbb{B}_{2N}$ is a ball of radius $1 - ||w||^2$ (in $2N-1$ space). For $w \in \mathbb{B}_{2N}$ and $1 \le i \le N$ define

$$n_i(w) = E_{2i-1} \quad \text{if} \quad |w_{2i}| + |w_{2i-1}| = 0$$

and

$$n_i(w) = \frac{\bar{w}_{2i}E_{2i-1} - \bar{w}_{2i-1}, E_{2i}}{||\bar{w}_{2i}E_{2i-1} - \bar{w}_{2i-1}E_{2i}||}$$

otherwise. These vector fields satisfy

$$||n_i(w)|| = 1$$

$$\langle n_i(w), n_k(w)\rangle = 0 \qquad i \neq k$$

$$\langle n_i(w), w\rangle = 0 \quad .$$

Low's observation is this. For $b > 0$ there is $\delta > 0$ such that: If I is a set of indices, $I \subset \{1,2,\ldots,N\}$ and w, w_i $(i \in I)$ are in $\mathbb{B}_{2N}$ with $|w - w_i| \le \delta$ and $|w|, |w_i| \ge b$ then there exist orthogonal vectors $n_i \in T_w$, $i \in I$, such that $|n_i - n_i(w)| < \varepsilon$. The initial induction step involves a close scrutiny of the function

$$g(z) = \sum_{i=1}^{N} g_i(z)$$

where

$$g_i(z) = \sum_{j=1}^{N_i} \alpha_{ij}p_{ij}(z)n_i(f(z_{ij})) \; .$$

The scalars

$$\alpha_{ij} = \left(\frac{1 - |f(z_{ij})|^2}{N}\right)^{1/2}$$

and the functions $p_{ij}(z)$ are scalar valued peaking functions

$$p_{ij}(z) = \exp(-m(1 - \langle z, z_{ij}\rangle)) \; .$$

Low has results similar to Theorem 5.1 for C^2 strictly pseudoconvex domains. We mention one more very interesting and pertinent result of this type. It is due to F. Forstneric [5] and although his results are valid for bounded strictly convex domains we refer only to the ball situation.

<u>Theorem</u> 5.2. <u>There is a positive integer N with the following property. If $h : \mathbb{B}_n \to C^p$ is a nonconstant holomorphic mapping such that $||h||^2$ extends as a continuous function to $\bar{\mathbb{B}}_n$ and $||h(z)|| \le a < 1$, then there is a holomorphic mapping $f : \mathbb{B}_n \to C^{2N}$ such that $F = (f,h) : \mathbb{B}_n \to C^{2N+p}$ maps $\mathbb{B}_n$ properly into $\mathbb{B}_{2N+p}$.</u>

A significant corollary is the following.

<u>Corollary</u> 5.3. <u>For each integer</u> $n \geq 1$ <u>there is a proper holomorphic embedding</u> $F : \mathbb{B}_n \to \mathbb{B}_{n+2+2N}$ <u>such that</u> F <u>can not be extended continuously to</u> $\overline{\mathbb{B}}_n$.

The techniques used to prove Theorem 5.2 are similar to those used by Low. For the corollary one produces two harmonic functions u_1 and u_2 on $\mathbb{B}_1$ which extend continuously to $\overline{\mathbb{B}}_1$ and such that the following holds. If v_j is the harmonic conjugate of u_j then

$$e^{2u_1} + e^{2u_1} = 1 \qquad \text{on} \qquad |z| = 1 ,$$

and v_j can not be extended continuously to $\overline{B_1}$. Then with $g_j = e^{u_j + iv_j}$ $j = 1,2$ define $G_j(z_1,\dots,z_n) = g_j(z_1)$ for $Z = (z_1,\dots,z_n) \in \mathbb{B}_n$. For $0 < \varepsilon < \frac{1}{2}$ the mapping $h(Z) = (Z, G_1(Z), G_2(Z))$ maps $\mathbb{B}_n \to \mathbb{B}_{n+2}$ and satisfies the hypothesis of Theorem 4.2. This establishes the corollary.

Finally we mention another result in this same direction. It is due to J. Globevnik [G].

<u>Proposition</u> 5.4. <u>Let</u> $n \geq 1$. <u>For every</u> $M \geq M(n)$ <u>there is a continuous mapping</u> $g : \overline{\mathbb{B}_n} \to \overline{\mathbb{B}_M}$ <u>which is holomorphic in</u> $\mathbb{B}_n$ <u>and satisfies</u> $g(\partial\mathbb{B}_n) = \partial\mathbb{B}_M$.

Let us close this section with some problems and observations. Is it possible to construct a proper mapping from $\mathbb{B}_n \to \mathbb{B}_{n+1}$ which can not be extended continuously to $\overline{\mathbb{B}_n}$? One expects that this is possible but we do not see a technique for approaching this question. If f is a proper holomorphic mapping from $\mathbb{B}_n \to \mathbb{B}_{n+1}$ which extends to $\overline{\mathbb{B}_n}$ as a Lipschitz mapping, is f holomorphic on $\overline{\mathbb{B}_n}$?

There is some contact between Lemma 4.4 and Corollary 5.3. If we choose $k = n + 1$ in Lemma 4.4 it is not difficult to use Lemma 4.3 to conclude that if f is in $I^2(n,n+1)$ then f is holomorphic on $\overline{\mathbb{B}_n}$. In fact this result is valid if f satisfies the C^2 smoothness criteria on an open dense subset of $\partial\mathbb{B}_n = S$. In Corollary 5.3 the only obvious discontinuities occur for the functions G_1 and G_2 on the curve $\{(e^{i\theta},0,\dots,0)\}$. Using the technique of [8] Professor Sibony has shown us a construction of a proper map from $\mathbb{B}_n \to \mathbb{B}_{2N}$ which can not be holomorphically continued over any open cap <u>in</u> S. The following question is probably unfair but it is of interest for examples. If $f(Z) = z_1^2(1 - z_2^2)^{-1/2}$ and $f_1(Z) = f(Z)/||f||_\infty$

do there exist functions f_2 and f_3 on $\mathbb{B}_2$ such that (f_1,f_2,f_3) is a proper map into $\mathbb{B}_3$?

6. The holomorphic continuation.

Earlier we showed that if $f \in I^2(2,3)$ then f was a rational mapping $\frac{p}{q}$. However, the following holds.

<u>Theorem</u> 6.1. <u>If</u> $f \in I^2(2,3)$ <u>then</u> f <u>is holomorphic on</u> $\overline{\mathbb{B}_2}$.

Proof. We know that $f = \frac{p}{q} = \frac{(p_1,p_2,p_3)}{q}$ where $\deg q \leq 2$ and degree $p_j \leq 3$. If there is $Z_0 \in \mathbb{B}_2$ such that $q(Z_0) = 0$ we can show that f is equivalent to one of Faran's maps. Without loss of generality we can assume $Z_0 = e = (1,0)$ and $q(e) = 0$. Thus

$$q(Z) = (1-z_1)(1-\alpha z_1-\beta z_2) + \gamma z_2^2$$

and $q \neq 0$ on $\mathbb{B}_2$. We establish the theorem for the special case $\beta = 0$. The general result (with $\beta \neq 0$) is similar and will appear in a forthcoming paper. Thus

$$q(Z) = (1 - z_1)(1 - z_1) + \gamma z_2^2$$

and $q \neq 0$ on $\mathbb{B}_2$. The mapping f has maximal rank off a finite set so we may assume $Df(0) = L$ is nonsingular. Also by composing f with a linear map (which leaves q unchanged) we may assume

$$L(z_1,z_2) = (a_{10}z_1 + a_{01}z_2, b_{01}z_2, 0)$$

where $a_{10} > 0$ and $b_{01} > 0$. With $\lambda \in C$, $|\lambda| = 1$ and $||Z|| = ||V|| = 1$, $\langle Z,V\rangle = 0$ we have

6.1.1 $$\langle f(Z + \lambda V), f(Z)\rangle \equiv 1 .$$

Write $P = (P_1,P_2,P_3) = L + Q + T$ where L,Q,T are homogeneous polynomial maps of degree 1, 2 and 3 respectively. If $Z = (z_1,z_2)$ and $V = (v_1,v_2) = (\bar{z}_2,-\bar{z}_1)$ then 6.1.1 implies (by equating coefficients of λ^3) that

$$0 = \langle T(V),P(Z)\rangle .$$

Hence, the inner product

6.1.2 $$\langle T(z_1,z_2),P(\bar{z}_2,-\bar{z}_1)\rangle$$

is holomorphic in C^2 and zero on S. Examining the fourth degree terms in 6.1.2 = 0 one concludes ($T = (T_1,T_2,T_3)$, etc.)

6.1.3 $$T_1(Z)(a_{10}z_2 - a_{01}z_1) - b_{01}z_1T_2(Z) = 0 .$$

There must exist homogeneous quadratics R_1 and R_2 with

$$T_1(Z) = z_1R_1(Z)$$

$$T_2(Z) = (a_{10}z_2 - \bar{a}_{01}z_1)R_2(Z) \ .$$

Further, 6.1.3 implies

$$R_1(Z) - b_{01}R_2(Z) = 0$$

so that

$$R_2(Z) = \frac{1}{b_{01}} R_1(Z) \ .$$

For a dense set of Z in S and $\lambda \in C$, $|\lambda| = 1$

6.1.4 $$\langle f(\lambda Z), f(\frac{1}{\bar{\lambda}} Z)\rangle \equiv 1 \ .$$

This implies

$$\langle P(\lambda Z), P(\frac{1}{\bar{\lambda}} Z)\rangle = S(\lambda Z)\overline{S(\frac{1}{\bar{\lambda}} Z)} \ .$$

Equating coefficients of λ^2 yields

$$T_1(Z)(a_{10}\bar{z}_1 + \bar{a}_{01}\bar{z}_2) + T_2(Z)(b_{01}\bar{z}_2) = \alpha z_1^2 + \gamma z_2^2 \ .$$

Using the relationship for R_1 and R_2 in this equation

$$R_1(Z)(a_{10}|z_1|^2 + \bar{a}_{01}z_1\bar{z}_2) + R_1(Z)(a_{10}|z_2|^2 - \bar{a}_{01}z_1\bar{z}_2) = \alpha z_1^2 + \gamma z_2^2$$

Thus

$$R_1(Z) = \frac{1}{a_{10}}(\alpha z_1^2 + \gamma z_2^2) \ .$$

We have now

$$T_1(Z) = \frac{z_1}{a_{10}}(\alpha z_1^2 + \gamma z_2^2)$$

$$T_2(Z) = (\frac{z_2}{b_{01}} - \frac{\bar{a}_{10}}{a_{10}b_{01}} z_2)(\alpha z_1^2 + \gamma z_2^2)$$

and using the continuity at e (for $\overline{\mathbb{B}_2}$) one sees

6.1.5 $$f_1(Z) = \frac{a_{10}z_1 + a_{01}z_2 + a_{20}z_1^2 + a_{11}z_1z_2 + a_{02}z_1^2 + \frac{z_1}{a_{10}}(\alpha z_1^2 + \gamma z_2^2)}{(1 - \alpha z_1)(1 - z_1) + \gamma z_2^2}$$

$$= \frac{z_1(a_{10} - \frac{\alpha}{a_{10}}z_1)(1-z_1) + a_{01}z_2(1-z_1) + z_2^2(a_{02} + \frac{\gamma}{a_{10}} z_1)}{q}$$

$$f_2(Z) = \frac{b_{01}z_2 + b_{20}z_1^2 + b_{11}z_1z_2 + b_{02}z_2^2 + (\frac{z_2}{b_{01}} - \frac{\bar{a}_{01}}{a_{10}b_{01}}z_1)(\alpha z_1^2 + \gamma z_z^2)}{q}$$

$$= \frac{z_2(b_{01} - \frac{\alpha}{b_{01}} z_1)(1-z_1) + \frac{\alpha\bar{a}_{01}}{a_{10}b_{01}} z_1^2(1-z_1) + z_2^2(b_{02} - \frac{\gamma\bar{a}_{01}}{a_{10}b_{01}} z_1) + \frac{\gamma}{b_{01}} z_2^3}{q}$$

$$f_3(z) = \frac{c_{20}z_1^2 + c_{11}z_1z_2 + c_{02}z_2^2 + c_{30}z_1^3 + c_{21}z_1^2z_2 + c_{12}z_1z_2^2 + c_{03}z_2^3}{q}$$

$$= \frac{c_{20}z_1^2(1-z_1) + c_{11}z_1z_2(1-z_1) + z_1^2(c_{02} + c_{12}z_1) + c_{03}z_2^3}{q}$$

For $|\phi| < \frac{\pi}{2}$ (a real number), ψ real and $0 < \varepsilon < 2\cos\phi$ define a curve $\Gamma(\varepsilon,\phi,\psi) = \Gamma$ in S by

$$z_1 = 1 - \varepsilon e^{i\phi}$$

$$z_2 = \sqrt{2\varepsilon\cos\phi - \varepsilon^2}\, e^{i\psi} .$$

Letting $r \to 1$ and using the continuity of f on $\mathbb{B}_2$ we obtain,

$$f_1(1,0) = \frac{a_{10} - \frac{\alpha}{a_{10}}}{1-\alpha} = \lim_{r\uparrow 1} f_1(r,0)$$

6.1.6 $$f_2(1,0) = \frac{a\,\bar{\alpha}_{01}}{a_{10}b_{01}(1-\alpha)} = \lim_{r\uparrow 1} f_2(r,0)$$

$$f_3(1,0) = \frac{c_{20}}{1-\alpha} = \lim_{r\uparrow 1} f_3(r,0)$$

By substituting the equations for $\Gamma(\varepsilon)$ into 6.1.5 and using 6.1.6 and letting $\varepsilon \to 0$ we deduce

$$\frac{(a_{10} - \frac{\alpha}{a_{10}})e^{i\phi} + (a_{02}\ \frac{\gamma}{a_{10}})(2\cos\phi)e^{2i\psi}}{(1-\alpha)e^{i\phi} + \gamma(2\cos\phi)e^{2i\psi}} = \frac{a_{10} - \frac{\alpha}{a_{10}}}{1-\alpha}$$

6.1.7 $$\frac{\frac{\alpha\,\bar{a}_{01}}{a_{10}b_{01}}e^{i\phi} + (b_{02} - \frac{\gamma\bar{a}_{01}}{a_{10}b_{01}})(2\cos\phi)e^{2i\psi}}{(1-\alpha)e^{i\phi} + \gamma(2\cos\phi)e^{2i\psi}} = \frac{\alpha\,\bar{a}_{01}}{a_{10}b_{01}(1-\alpha)}$$

$$\frac{c_{20}e^{i\phi} + (c_{02} + c_{12})(2\cos\phi)e^{2i\psi}}{(1-\alpha)e^{i\phi} + \gamma(2\cos\phi)e^{2i\psi}} = \frac{c_{20}}{1-\alpha}$$

Hence

$$a_{02} = -\frac{\gamma}{a_{10}} + \frac{\gamma(a_{10} - \frac{\alpha}{a_{10}})}{1-\alpha}$$

$$b_{02} = \frac{\gamma\bar{a}_{01}}{a_{10}b_{01}} + \frac{\alpha\bar{a}_{01}\gamma}{a_{10}b_{01}(1-\alpha)}$$

$$c_{02} = -c_{12} + \frac{c_{20}\gamma}{1-\alpha} \quad .$$

Using this information we can rewrite equation 6.1.5 as follows

$$f_1(Z) = \frac{z_1(a_{10} - \frac{\alpha}{a_{10}}z_1)(1-z_1) + \gamma z_2^2(\frac{a_{10} - \frac{\alpha}{a_{10}}}{1-})a_{01}z_2(1-z_1) + \frac{\gamma}{a_{10}}z_2^2(1-z_1)}{q(Z)}$$

$$f_2(Z) = \frac{\frac{\alpha\bar{a}_{01}}{a_{01}b_{01}}(z_1^2(1-z_1) + \frac{\gamma}{(1-\alpha)}z_2^2) + z_2(1-z_1)(b_{01} - \frac{\alpha}{b_{01}}z) + z_2^2(1-z_1)\cdot}{q(Z)}$$

$$\left(\frac{\gamma\bar{a}_{01}}{a_{10}b_{01}}\right)\left(\frac{\gamma}{b_{01}}z_2^3\right)$$

$$f_3(Z) = \frac{c_{20}(z_1^2(1-z_1) + \frac{\gamma}{1-\alpha}z_2^2) + c_{11}z_1z_2(1-z_1) - c_{12}z_2^2(1-z_1) + c_{03}z_1^3}{q(Z)}$$

Since f is $C^1(\overline{\mathbb{B}_2})$ we have

$$\frac{f(e) - f(Z)}{||e - z||} \longrightarrow Df(e)(0,e^{i\psi})$$

as $\varepsilon \to 0$, where

$$Z = (z_1,z_2) = (1 - \varepsilon e^{i\phi}, \sqrt{2\varepsilon\cos\phi - \varepsilon^2}\; e^{i\psi}).$$

Since

$$f_1(1,0)-f_1(z_1,z_2) = \frac{(1-z_1)^2(\frac{a_{10} - \frac{\alpha}{a_{10}}}{1-\alpha} - \frac{\alpha}{a_{10}}z_1) - a_{01}z_2(1-z_1) + \frac{\gamma}{a_{10}}z^2(1-z_1)}{q(Z)}$$

and

$$||(1,0) - (z_1,z_2)||^2 = 2\varepsilon\cos\phi$$

we conclude

$$e^{-i\psi}D_1f(e)(0,e^{i\psi}) = Df_1(e)(0,1) = \frac{-a_{01}e^{i\phi}}{(1-\alpha)e^{i\phi} + \gamma e^{2i\psi}(2\cos\phi)} =$$

$$= \frac{-a_{01}}{(1-\alpha) + \gamma(2\cos\phi)e^{i(2\Psi-\phi)}} .$$

Therefor either $\gamma = 0$ or $a_{01} = 0$. If $\gamma = 0$ the numerators of f_1 and f_2 must have $(1-z_1)$ as a factor. Thus f_1 and f_2 reduce to a quadratic over a linear factor. The function f_3 would have the value

$$f_3(Z) = \frac{c_{20}z_1^2 + c_{11}z_1z_2 - c_{12}z_2^2}{1-\alpha z_1} + \frac{c_{03}z_2^3}{(1-\alpha z_1)(1-z_1)} .$$

Both functions on the right of this equation are C^1 functions on $\overline{\mathbb{B}_2}$. This means

$$\frac{e^{i\psi}c_{03}(2\varepsilon\cos\phi - \varepsilon^2)^{3/2}e^{3i\Psi}}{[(1-\alpha(1-\varepsilon e^{i\phi})(\varepsilon e^{i\phi})]\sqrt{2\varepsilon\cos\phi - \varepsilon^2}}$$

has a limit as $\varepsilon \to 0$ and this limit is constant. Therefore $c_{03} = 0$ and we are reduced to $f = \frac{p}{S}$ where p is quadratic and $S(Z) = (1-z_1)$ is linear. We have

$$f_1(Z) = \frac{z_1(a_{10} - \frac{\alpha}{a_{10}}z_1) + a_{01}z_2}{1-\alpha z_1}$$

$$f_2(Z) = \frac{z_2(b_{01} - \frac{\alpha}{b_{01}}z_1) + \frac{\alpha\bar{a}_{01}}{a_{10}b_{01}}z_1^2}{1-\alpha z_1}$$

$$f_3(Z) = \frac{c_{20}z_1^2 + c_{11}z_1z_2 - c_{12}z_2^2}{1-\alpha z_1}$$

If $|\alpha| = 1$, then clearly $a_{01} = 0$, $a_{10} = 1$, $b_{01} = 1$ and

$$c_{20} = c_{11} = c_{12} = 0 .$$

This proves f is analytic on $\overline{\mathbb{B}_2}$ (and if $|\alpha| > 1$ there is nothing to prove).

Returning to the case $\gamma \neq 0$ and $a_{01} = 0$ we have

$$f_2(Z) = \frac{z_2(1-z_1)(b_{01} - \frac{\alpha}{b_{01}}z_1) + \frac{\gamma}{b_{01}}z_2^3}{(1-\alpha z_1)(1-z_1) + \gamma z_2^2} .$$

So $f_2(e) = 0 = \lim_{r\to 1} f_2(r,0)$. Then

$$Df_2(e)(0,1) = \frac{e^{i(\phi+\Psi)}\left(b_{01} - \frac{\alpha}{b_{01}}\right) + \frac{\gamma}{b_{01}}(2\cos\phi)e^{3i\Psi}}{(1-\alpha)e^{i\phi} + \gamma(2\cos\phi)e^{2i\Psi}}$$

and as a consequence

$$\frac{\left(b_{01} - \frac{\alpha}{b_{01}}\right) + \frac{\gamma}{b_{01}}(2\cos\phi)e^{i(2\Psi-\phi)}}{(1-\alpha) + \gamma(2\cos\phi)e^{i(2\Psi-\phi)}}$$

is a constant. Since $\gamma \neq 0$ we see that

$$\left(b_{01} - \frac{\alpha}{b_{01}}\right) - c(1-\alpha) = \gamma e^{i(2\Psi-\phi)}(2\cos\phi)\left(c - \frac{1}{b_{01}}\right) .$$

Hence, $c = \frac{1}{b_{01}}$ and

$$b_{01} - \frac{\alpha}{b_{01}} - \frac{1}{b_{01}} + \frac{\alpha}{b_{01}} = 0 .$$

Consequently $b_{01} = 1$ and we conclude

$$f_2(Z) \equiv z_2 .$$

On the slice $(0,z_2)$ we have

$$|f_1(0,z_2)|^2 + |f_3(0,z_2)|^2 \leq 1 - |z_2|^2 .$$

An application of the maximal principle shows that

$$f_1(0,z_2) \equiv 0 \qquad |z_2| \equiv 1$$

$$f_3(0,z_2) \equiv 0 \qquad |z_2| \equiv 1 .$$

Recalling the definition of f_1 and the fact that $a_{01} = 0$ yields

$$f_1(0,z_2) = \frac{\gamma z_2^2\left|\left(\frac{a_{10} - \frac{\alpha}{a_{10}}}{1-\alpha}\right) - \frac{\gamma}{a_{10}}\right|}{1 + \gamma z_2^2} \equiv 0 .$$

This yields (since $\gamma \neq 0$)

$$a_{10} - \frac{\alpha}{a_{10}} = \frac{1}{a_{10}} - \frac{\alpha}{a_{10}} \quad \text{and} \quad a_{10} = 1.$$

Finally, returning to the definition of f_1 we have

$$f_1(Z) \equiv z_1$$

$$f_3(Z) \equiv 0 .$$

This concludes the proof.

Notice that we proved the following. If $f(Z) = \frac{P(Z)}{q(Z)}$ is a proper, rational mapping of $\mathbb{B}_2 \to \mathbb{B}_3$ with $\deg P_j \leq 3$, $j = 1,2,3$ and $\deg q \leq 2$ and if f is $C^1(\overline{\mathbb{B}_2})$ then f is holomorphic on $\overline{\mathbb{B}_2}$. The assumption that $f \in I^2(2,3)$ was used only to show that f had the desired rational form.

7. Resume and questions.

The results to date on proper mappings of balls in C^n is impressive. There are at present contrasts in comparing the works [2], [3], [12] and those of [5], [6] and [11]. At present those contrasts center about the fact that increasing the dimension of the ball in the range allows the construction of proper maps that behave badly near $\partial\mathbb{B}_n = S$ and for the seemingly interesting case of maps $\mathbb{B}_n \to \mathbb{B}_{n+1}$ we have produced only smooth maps on $\overline{\mathbb{B}_n}$. The contrast may only be illusory. The second and more positive information shows that certain smoothness assumptions for proper maps on $\overline{\mathbb{B}_n}$ guarantees holomorphy on the closed ball $\overline{\mathbb{B}_n}$. Perhaps the most significant and pertinent question was posed at the end of Section 5. Namely to produce "pathological" proper mappings of $\mathbb{B}_n$ into $\mathbb{B}_{n+1}$. Our results indicate that the C^1 condition may be the critical case.

One can also pose the question as to which proper maps from $\mathbb{B}_n \to \mathbb{B}_{n+1}$ have finite volume. Considering the case $n = 2$ and

$$dM = \left| \left(\sum_{j=1}^{3} \left| \frac{\partial f_j}{\partial z_1} \right|^2 \right) \left(\sum_{j=1}^{3} \left| \frac{\partial f_j}{\partial z_2} \right|^2 \right) - \left| \sum_{j=1}^{3} \frac{\partial f_j}{\partial z_1} \left(\overline{\frac{\partial f_j}{\partial z_2}} \right) \right|^2 \right| dV$$

where dV is Lebesgue measure on $\mathbb{R}^4$ we have the following question. For which proper mappings of $\mathbb{B}_2 \to \mathbb{B}_3$ is

$$\int_{\mathbb{B}_2} dM < +\infty \ ?$$

It is known that there exist functions $f_1(Z)$ on the ball in C^n which are in the ball algebra (continuous on $\overline{B_n}$) and also satisfy a Lip α condition on $\overline{\mathbb{B}_n}$. Such functions can be produced which satisfy

$$|f_1(Z)| \leq 1 \quad \text{and}$$
$$\mu(z \in \partial\mathbb{B}_n = S \mid |f_1(Z)| = 1) > 0,$$

where μ is Lebesgue measure on S. Clearly such functions can not be a coordinate in a proper map from $\mathbb{B}_n$ into $\mathbb{B}_{n+1}$. The question is if $g(Z) = (g_1(z),\ldots,g_{n+1}(Z))$ is a proper continuous mapping $\overline{\mathbb{B}_n} \Rightarrow \overline{\mathbb{B}_{n+1}}$, which also satisfies a Lip condition, what measure theoretic and topological properties can the level sets of g_j

$$U_j(P) = \{Z \in S \mid |g_j(Z)| = P \qquad 0 \leq P \leq 1$$

have?

As a final and peripheral point we mention the following. If the reader will compare example (3.1) and (3.3) they will see a significant difference. That is (3.1) is a "planar" mapping and (3.3) is not. We conjectured that if f is in $I^2(n,k)$ with $n < k \leq 2n - 2$ then f is equivalent to (3.1). Under the assumption that f is analytic on $\overline{\mathbb{B}_n}$ Faran [4] has proven this equivalence. Thus even with certain smoothness assumptions one can see differences in the class of proper mappings from $\mathbb{B}_n \to \mathbb{B}_k$ if f is large. The number $2n-1$ is interesting to us. There is a paper by H. Whitney [13] which discusses singularities of smooth k manifolds in $(2k-1)$ space. In complex space with $k = 2n-1$ and $2k-1 = 4n-3$ these are the real dimensions of the balls in C^n and C^{2n-1}. In presenting some of the material in seminar it was pointed out to one of the authors (by Professors J. Damon and C.T.C. Wall) that some of these examples of proper mappings occur in singularity theory. In particular for $n \leq k \leq 2n - 2$ the mappings

$$(z_1,\ldots,z_n) \to (z_1,\ldots,z_n, 0,\ldots,0)$$

occur since there are no other stable models for these indices. They were also aware of example (3.3) and suggested that it arises in unfolding (and stabilizing) the example given by Whitney (the Whitney umbrella). Although there may not be a deep connection between the topics it would be interesting to know if other examples occuring in singularity theory might not be useful to obtain more examples of proper mappings.

References

[1] Cima, J., Krantz, S.G. and Suffridge, T.: A reflection principle for proper holomorphic mappings of strongly pseudoconvex domains and applications. Math. Zeit. 186, 1-8 (1984).

[2] Cima, J. and Suffridge, T.: A reflection principle with applications to proper holomorphic mappings. Math. Ann. 265, 189-500 (1983).

[3] Faran, J.: Maps from the two-ball to the three-ball and maps taking lines to plane curves. Invent. Math. 68 441-475 (1982).

[4] Faran, J. J.: The linearity of proper holomorphic maps between balls in the low codimension case. Preprint.

[5] Forstneric, F.: Embedding strictly pseudoconvex domains into balls. T.A.M.S., 295, 347-367 (1986).

[6] Globevnik, J.: Boundary interpolation by proper holomorphic maps". Preprint.

[7] Globevnik, J. and Stout, E.L.: Boundary regularity for holomorphic maps from the disc to the ball. Preprint.

[8] Hakim, M. and Sibony, N.: Fonctions holomorphes bornees sur la boule unite de C^n. Invent. Math. 67, 213-222 (1982).

[9] Lempert, L.: Imbedding strictly pseudoconvex domains into balls. Amer. J. Math. 104, 901-904 (1982).

[10] Lewy, H.: On the boundary behavior of holomorphic mappings. Acad. Naz. Lincei 35, 1-8 (1977).

[11] Low, E.: Embeddings and proper holomorphic maps of strictly pseudoconvex domains into polydiscs and balls. Preprint.

[12] Webster, S.: On mapping an n-ball into an (n+1)-ball in complex space. Pacific J. Math. 81, 267-272 (1979).

[13] Whitney, H.: Singularities of smooth k manifolds in (2k-1) space. Annals of Math. 45, 220-247 (1944).

Finite-Type Conditions for Real Hypersurfaces in $\mathbb{C}^n$

John P. D'Angelo*

University of Illinois
Urbana, Illinois 61801

Introduction

Complex function theory in several variables requires a thorough study of the influence of the geometry of the boundary of a domain on the domain itself. Let us suppose that Ω is an open domain in $\mathbb{C}^n$ and that its boundary M is a smooth real submanifold of $\mathbb{C}^n$. One measures algebraic-geometric or differential-geometric information on M and uses this to derive consequences for the function theory on Ω. In this paper we organize and survey those geometric conditions that play a crucial role in case M has a degenerate Levi form.

There are several distinct concepts that go by the name "point of finite type". These have arisen since 1972, when Kohn [K1] first defined the concept for points on the boundaries of smoothly bounded pseudoconvex domains in $\mathbb{C}^2$. He established that this notion was a sufficient condition for subelliptic estimates in the $\bar{\partial}$-Neumann problem. In 1974, Greiner [G] established the necessity of this condition for the estimates in this case. There are many conceivable generalizations to higher dimensions. In this paper, we fit into one framework most of these possibilities. We say that p is a point of finite 1-type if the order of contact of all complex analytic varieties with M at p is bounded. Catlin [C1, C2] has proved that this condition is necessary and sufficient for the subelliptic estimates on $(0,1)$ forms, in the case where Ω is a smoothly bounded pseudoconvex domain. More generally, we study in section I the notions of finite q-type and finite regular q-type. These involve the orders of contact of q dimensional (perhaps singular) complex analytic varieties and q dimensional complex manifolds, respectively. It then becomes clear that one wishes to assign numerical invariants, for each integer between 1 and $n-1$, that describe the geometry of M. In case that Ω is a smoothly bounded pseudoconvex domain, and each boundary point is a point of finite type, it follows from the subelliptic estimates and the work of Bell [Be1] that these invariants will actually be biholomorphic invariants of the domain itself. This is because of Bell's result that subelliptic estimates imply that a biholomorphism of such domains extends to be a diffeomorphism of the boundaries.

We consider here numerical invariants $S_q(M,p)$ of a real hypersurface M containing p, where q is an integer between 1 and $n-1$. We want these

* Partially supported by the NSF Grants MCS-8108814 (A04) and DMS-8501008 and by The Institute for Advanced Study.

invariants to be intrinsic to M and independent of local biholomorphic coordinate changes. We also want them to measure, at least in the pseudoconvex case, the degeneracy of the Levi form. There are many choices.

In section I we consider S_q to be the maximum order of contact of q dimensional complex analytic subvarieties of $\mathbb{C}^n$, or to be the maximum order of tangency of q dimensional complex analytic submanifolds of $\mathbb{C}^n$. These are distinct notions. In section II we consider intersection multiplicities for the S_q. These numbers have important semi-continuity properties. In section III we consider some choices that arise from generalizing Kohn's original definition using iterated commutators of (1,0) tangent vector fields. In Section IV we discuss the relationship of these ideas to subelliptic estimates and Catlin's multitype. In section V we consider another notion of finiteness, due to Baouendi-Treves-Jacobowitz [BJT], that arises in the analyticity of CR mappings. The final section contains a list of open questions.

This paper has many examples and references and some proofs. The results on intersection multiplicities in section II and the general organization of the material is new. It seems to the author that the idea of assigning only numerical invariants to the boundary is too naive; one should assign objects such as families of ideals of holomorphic functions to each boundary point. Perhaps the methods of algebraic geometry will be useful in attempting to give a complete list of invariants, a problem beyond the scope of this article.

The author acknowledges the participants of the international conference on partial differential equations in complex analysis held in Albany, 1985, and of the complex analysis week held at Penn State in March, 1986. He particularly thanks the organizers, Michael Range and Steve Krantz, respectively, for their encouragement in preparing such an article. Finally, he also acknowledges the hospitality of The Institute for Advanced Study, where he wrote this article.

I. Points of finite q-type and finite regular q-type

Kohn first introduced the notion of point of finite type on a pseudoconvex real hypersurface in the space of two complex variables. He [K1] was able to establish the sufficiency of this condition for subelliptic estimates in the $\bar{\partial}$-Neumann problem. When Greiner [G1] established the necessity of this condition for the estimates, it became clear that the notion of point of finite type was of basic importance in the theory of functions of several complex variables. The generalization to higher dimensions is not obvious. In fact, several different concepts are appropriate for several different problems. Many different definitions appear in the literature.

In this section we describe how many of these definitions fit into one algebraic-geometric framework. It turns out that one describes the geometry of a real hypersurface M in $\mathbb{C}^n$ by analyzing how closely ambient complex analytic varieties contact M . Many of the interesting phenomena arise because of the necessity of considering singular varieties. This difficulty does not arise in two complex variables, and thus all the notions turn out to be equivalent in this case.

Let (M,p) denote the germ at p of a real hypersurface in $\mathbb{C}^n$. Let (V,p) denote the germ of p of a complex analytic subvariety of $\mathbb{C}^n$. We will write $\mathscr{I}(M)$ and $\mathscr{I}(V)$ for the ideals of germs of functions vanishing on M and V , where the relevant rings are the smooth and holomorphic germs, respectively. Suppose that (V,p) is the germ of an analytic subvariety. We can always find a non-constant germ of a holomorphic map

1. $z : (\mathbb{C},0) \longrightarrow (V,p)$

because any irreducible one dimensional subvariety of V has a normalization. To measure the contact of V with M at p , we pull back to such curves, and measure the order of vanishing. Since V can be singular, we must divide by the multiplicity of z , the order of the singularity of this one dimensional branch. This leads to the notion of point of finite type (which we will call "point of finite 1-type"). Before making the definition, we need some notation. We let $v(z)$ denote the multiplicity of the map z , and $v(z^* r)$ denote the order of vanishing of the pull-back map given by composition. Here r is any smooth real valued function.

2. Definition. [D 1] . A point p on a real hypersurface M is called <u>a point of finite 1-type</u> if there is a constant C so that

$$v(z^* r)/v(z) \le C ,$$

whenever z is a non-constant holomorphic germ as in 1, and r is a defining function for M . The infimum of all such constants C is called the 1-type of p , and is denoted by $\Delta(M,p)$ or $\Delta_1(M,p)$. It is easy to verify that this condition is independent of the defining function.

Catlin [C 3] has proved that finite 1-type is necessary and sufficient for a subelliptic estimate on $(0,1)$ forms, in case M is the boundary of a smoothly bounded pseudoconvex domain. Before proceeding to the more general framework, we state a simple proposition.

3. Proposition. $\Delta(M,p)$ can also be expressed as

$$\sup \sup \{a \in \mathbb{R}^+ : \lim (\mathrm{dist}(z,M)/|z-p|^a \text{ exists}\} .$$

Here the first supremum is taken over all one dimensional complex analytic varieties, the second is taken over a , and the limit is taken as z tend to p while lying in V .

The function $\Delta(M,p)$ is not semi-continuous from either side, although it is locally bounded [D1,2]. To prove the local boundedness seems to require a certain amount of algebraic geometry, especially the notion of intersection multiplicity.

This is the motivation for section II of this paper. Before turning to these ideas, we begin a general discussion of finite type conditions, by listing the properties we would like numerical invariants of M to have. We then proceed to a large number of possible definitions that satisfy only some of the properties. Only the multiplicities of section II will satisfy all the properties 4 (including upper semi-continuity) below.

4. Desired properties of numerical invariants. Let (M,p) be the germ of a real hypersurface in $\mathbb{C}^n$. We wish to assign numerical invariants $S_q(M,p)$, for each integer q between one and $n-1$, to the germ (M,p) that satisfy the following:

4.0 $S_q(M,p)$ is a positive real number or plus infinity.

4.1 $S_{n-1}(M,p) \leq S_{n-2}(M,p) \leq \dots \leq S_1(M,p)$.

4.2 $S_q(M,p) = 2$ if the Levi form has (at least) $n-q$ eigenvalues of the same sign at p. In particular, all the numbers equal 2 when M is strongly pseudoconvex from one side at p. (See section 3 for the definition of the Levi form.)

4.3 $S_q(M,p)$ is an invariant; this number does not depend on a choice of local coordinates or on a local defining equation for M.

4.4 $S_q(M,p)$ is a locally bounded function of p. (Even better would be an upper semi-continuous function of p.)

4.5 $S_q(M,p)$ is finitely determined. Suppose that M is defined by r in $\mathscr{I}(M)$ and that $S_q(M,p)$ is finite. Then there is an integer k so that $S_q(M,p) = S_q(M',p)$ whenever M' is defined by any r' that has the same k jet as r at p.

Finally, one hopes that these numbers have an intuitive geometric definition and that they arise in several applications. Many of our candidates will not satisfy 4.4 and 4.5. In order to obtain these for orders of contact, we must consider singular varieties. We recall a definition from [D 1], in a slightly more general context. Let Q denote any of the following rings of germs of smooth functions at p: holomorphic germs, smooth germs, real analytic germs, formal power series in the z's, or formal power series in the z's and the $\bar{z}$'s.

5. Definition. Let $\mathscr{I}$ be an ideal in Q. We put $\tau^*(\mathscr{I}) = \sup \inf v(z^*g)/v(z)$, where the infimum is taken over g in $\mathscr{I}$, and the supremum is taken over all non-constant holomorphic maps z as in definition 1.

6. Definition. The invariant $\Delta_q(M,p)$, a measure of the maximum order of contact of q dimensional varieties with M at p, is defined as follows:

$$\Delta_q(M,p) = \inf \tau^*((\mathscr{I}, w_1, \dots, w_{q-1}))$$

Here $\mathcal{I}$ is the ideal $\mathcal{I}(M)$, and the ideal in question is the ideal generated by $\mathcal{I}$ and q-1 linear forms at p . The infimum is taken over all such choices of linear forms. Note that, when q equals one, the definitions 2 and 6 are equivalent.

<u>7. Definition</u>. The invariant $T_q(M,p)$ is defined as in Proposition 3, with the supremum being taken over all q dimensional complex analytic varieties. The invariant $\Delta^{reg}{}_q(M,p)$ is defined as in Proposition 3, with the supremum being taken over q dimensional complex analytic manifolds. Alternative notations for $\Delta^{reg}{}_q(M,p)$ are reg $O^q(M,p)$ in [K 2] and $a^q(M,p)$ in [B 1].

<u>8. Remarks</u>. It is easy to verify that all the invariants of definitions 6 and 7 satisfy properties 4.1, 4.2, and 4.3. It is also clear that $\Delta^{reg} \leq T$. However, these numbers are not finite simultaneously. See Example 10 below. In Section III we see that when q equals n-1 , all the invariants give the same values. This also explains why there is only one viable concept in two complex variables.

<u>9. Definitions</u>. We say that p in M is a point of finite regular q-type if $\Delta^{reg}{}_q(M,p)$ is finite, and that p is a point of finite q-type if $\Delta_q(M,p)$ is finite.

<u>10. Example</u>. Let M be defined by the equation

$$r(z,\bar{z}) = 2\mathrm{Re}(z_4) + |z_1^2 - z_2^3|^2 + |z_3|^{2m} \; ; \quad m \geq 4 .$$

Let p denote the origin. We have the following values:

10.1 $\Delta^{reg}{}_1(M,p) = 2m$; given by $z_1 = z_2 = z_2 = 0$.

$T_1(M,p) = \Delta_1(M,p) = \infty$; given by $z_1^2 - z_2^3 = z_3 = z_4 = 0$.

10.2 $\Delta^{reg}{}_2(M,p) = 6$; given by $z_1 = z_4 = 0$.

$\Delta_2(M,p) = 2m$; given by $z_1^2 - z_2^3 = z_4 = 0$.

10.3 $\Delta^{reg}{}_3(M,p) = \Delta_3(M,p) = 4$; given by $z_4 = 0$.

We will return to this example when we discuss iterated commutators in Section III.

We now describe some of the properties of the function $\Delta_q(M,p)$.

<u>11. Proposition</u>. $T_q(M,p) \leq \Delta_q(M,p)$.

<u>Proof</u>. Let N denote $T_q(M,p)$. From its definition we can find, for any positive ε , a q-dimensional variety V for which

11.1 $\mathrm{dist}(z,M) \leq \mathrm{const}\,|z-p|^a$.

whenever z is close to p and lies in V , and $N-\varepsilon < a < N$. If we choose q-1 linear forms, the variety V' defined by $\mathcal{I}(V)$ and these linear forms must be at least one dimensional. We can therefore find a map z as in Proposition 1 whose image lies in V' . The inequality 11.1 holds along the image of z . Write w for the linear imbedding from $\mathbb{C}^{n-q+1}$ into $\mathbb{C}^n$ given by these linear forms. Letting t tend to zero, and writing ζ for z^*w , we obtain that

$$|r(\zeta(t))| \leq \text{const } |t|^{av(\zeta)} ,$$

and hence that $v(\zeta^* r)/v(\zeta) \geq a \geq N-\varepsilon$. This implies the desired result.

12. Theorem [D1]. Let M be a smooth real hypersurface of $\mathbb{C}^n$. Then the set of points of finite q-type is an open subset of M . In fact, the function $\Delta_q(M,p)$ is locally bounded.

13. Theorem. Let M be a real analytic hypersurface of $\mathbb{C}^n$. Then p is a point of finite q-type if and only if there is no germ of a complex analytic q dimensional subvariety of $\mathbb{C}^n$ containing p and lying in M .

14. Proposition. The function $\Delta_q(M,p)$ is finitely determined in the sense of property 4.5.

Proof. If $\Delta_q(M,p)$ is finite, choose any integer k larger than $C = \Delta_q(M,p)$. Suppose that r' is any smooth function with the same k-jet as r at p , where r is any defining function for M . The infimum in definition 6 is clearly attained for a generic w . For such a w , and any nonconstant holomorphic z , we have

14.1 $z^* w^* r' = z^* w^* r + z^* w^* (r-r')$

The second term on the right of 14.1 vanishes to order at least $(k+1)v(z)$, while the first vanishes to order at most $kv(z)$ by the choice of k . Therefore, the left side vanishes also to the same order as the first term on the right. Hence $v(z^* w^* r')/v(z) \leq C$, so property 4.5 is satisfied.

15. Corollary. The function $\Delta_q(M,p)$ satisfies all the properties 4, although it fails to be upper semi-continuous.

Proof. The failure of semi-continuity can be seen in example II.16. We have proved 4.5 in proposition 14, and stated 4.3 in Theorem 12. The property 4.1 follows trivially from the definition of $\Delta_q(M,p)$, and the properties 4.2 and 4.3 are safely left to the reader.

If we write M_k for the hypersurface defined by the k-jet of r at p , we see from Proposition 14 that, in the finite q-type case, $\Delta_q(M_k,p)$ eventually stabilizes to $\Delta_q(M,p)$. Here is a simple example from [D1].

16. Example. Put $r(z) = 2\mathrm{Re}(z_3) + |z_1^2 - z_2^3|^2 + |z_1|^8 - |z_2|^{12} + |z_1|^{2m}$.

Suppose that $m \geq 7$ and that p is the origin. Then

$$\begin{aligned}
\Delta_1(M_k,p) &= \infty && \text{for } 0 \leq k \leq 7 \\
&= 12 && \text{for } 8 \leq k \leq 11 \\
&= \infty && \text{for } 12 \leq k \leq 2m-1 \\
&= 3m && \text{for } 2m \leq k \leq \infty
\end{aligned}$$

We complete this section by noting that properties 4.4 and 4.5 fail to hold for the numbers $\Delta^{reg}_q(M,p)$. The 6 jet of example 16 furnishes us with an example where both of these properties fail when q equals one.

II. Multiplicities

There are many possible ways to measure a singularity; intersection multiplicities are one of the nicest. In this section we show how to define such numbers on a real hypersurface. This yields a collection of numbers $B_q(M,p)$ that satisfy all the properties of Section I. Before proceeding to the necessary algebra, the following example compares these numbers to the ones we have considered thus far.

Example 1. Put $r(z,\bar{z}) = 2\mathrm{Re}(z_n) + \sum^{n-1} |z_j|^{2m_j}$; let p denote the origin, and suppose that $m_1 \geq m_2 \geq \cdots \geq m_{n-1}$.

(1.1) $\Delta_q(M,p) = \Delta_q^{\mathrm{reg}}(M,p) = 2m_q$

(1.2) $B_q(M,p) = 2\prod_q^{n-1} m_j$.

In example 1, the collections of numbers (1.1) and (1.2) convey the same information. However, the number $B_1(M,p)$ includes all the information. It is this number that the author feels is the most useful. To define it, we need some basic formal algebraic notions.

Notation 2. Let p be a point in $\mathbb{C}^n$. We consider several local rings at p . $\mathcal{O}$ denotes the holomorphic germs, $\mathcal{A}$ denotes the real analytic, real valued germs, $\hat{\mathcal{O}}$ denotes the formal power series in $z-p$, and $\hat{\mathcal{A}}$ denotes the formally real formal power series in $(z-p, \overline{z-p})$. We say that $\sum C_{ab}(z-p)^a\overline{(z-p)}^b$ is formally real if

$$C_{ab} = \overline{C_{ba}} \quad .$$

Note that the Taylor series of a defining function for a smooth real hypersurface containing p gives us an element of $\hat{\mathcal{A}}$. Our next computation shows how to write this in terms of $\hat{\mathcal{O}}$. In case the defining function is an element of $\mathcal{A}$, the result will be in terms of $\mathcal{O}$.

Computation 3. Put $w = \sum C_{ab}(p)(z-p)^a\overline{(z-p)}^b$; assume that $C_{00}(p) = 0$ and that $C_{ab}(p) = \overline{C_{ba}}(p)$. We define elements of $\hat{\mathcal{O}}$ as follows:

(3.1) $h^p(z) = 4\sum C_{a0}(z-p)^a$

(3.2) $f_b^p(z) = (z-p)^b + \sum C_{ab}(p)(z-p)^a$

(3.3) $g_b^p(z) = (z-p)^b - \sum C_{ab}(p)(z-p)^a$.

Then we have

(3.4) $4w = 2\mathrm{Re}(h^p(z)) + \|f^p(z)\|^2 - \|g^p(z)\|^2$.

Note that, if w lies in $\mathcal{A}$, h^p , f_b^p , g_b^p are all elements of $\mathcal{O}$. In that case, according to [D 1, D 3] , the only complex subvarieties of $\mathbb{C}^n$ that can lie in the zero set of w must be defined by the equations (4.1) and (4.2):

(4.1) $h^p = 0$

(4.2) $f^p = U g^p$ or $f_b^p = \Sigma U_{bk} g_k^p$.

Here U is a unitary matrix of constants. This motivates the definition of the ideals $\mathcal{I}(U,p)$ in $\hat{\mathcal{O}}$ defined by

Definition 5. $\mathcal{I}(U,p)$ is generated by h^p and $f_b^p - \sum_k U_{bk} g_k^p$. Note that $\mathcal{I}(U,p)$ is a proper ideal in $\hat{\mathcal{O}}$ (or $\mathcal{O}$ in case w is in $\mathcal{A}$).

These ideals are the obstructions to finding complex analytic subvarieties in a real hypersurface. In other words, we have the following restatement of a result from [D 1].

Theorem 6. Let M be a real analytic hypersurface of $\mathbb{C}^n$, containing p . Let w be a defining function for M , and let $\mathcal{I}(U,p)$ be the ideals of definition 5. Then

(6.1) $\Delta_q(M,p)$ is finite $\Longleftrightarrow$ $\dim V(\mathcal{I}(U,p)) < q$ (for all U)

Now, in case $\dim V(\mathcal{I}(U,p))$ equals 0 for all U , there is a simple way to measure its singularity. We recall some analytic geometry [S,D 1].

Definition 7. Let $\mathcal{I}$ be a proper ideal in $\hat{\mathcal{O}}$ or $\mathcal{O}$. Its multiplicity, $D(\mathcal{I})$, is the dimension of the complex vector space $\hat{\mathcal{O}}/\mathcal{I}$, or $\mathcal{O}/\mathcal{I}$.

Theorem 8. (Nullstellensatz). Let $\mathcal{I}$ be a proper ideal in $\mathcal{O}$. The following are equivalent.

(8.1) $D(\mathcal{I}) < \infty$

(8.2) $V(\mathcal{I}) = \{p\}$ (or, $\dim V(\mathcal{I}) = 0$).

(8.3) There is an integer s so that ζ^s lies in $\mathcal{I}$ for every ζ that vanishes at p (s is independent of ζ)

Remark 9. In case $\mathcal{I}$ is defined by specific generators whose coefficients depend continuously on some parameter, then $D(\mathcal{I})$ depends upper semi-continuously on this parameter.

Example 10. Suppose $f:(\mathbb{C}^n,0) \longrightarrow (\mathbb{C}^n,0)$ is holomorphic, and $\mathcal{I}$ denotes the ideal generated by the components of f . Then $D(\mathcal{I})$ is the topological degree of f , i.e., its winding number about 0. It also has the interpretation as the generic number of roots to the equation $f(z) = w$ for w close to 0 . If the coefficients of f depend on some parameter ε , we have that $D(\mathcal{I}_\varepsilon) \le D(\mathcal{I}_0)$. For example, if $\mathcal{I}_\varepsilon = (\varepsilon z + z^2)$, we have $1 = D(\mathcal{I}_\varepsilon) \le D(\mathcal{I}_0) = 2$.

We extend the notion of multiplicities to varieties of positive dimension as follows. Suppose that $\dim(V(\mathcal{I})) = q$. Then, for a generic choice of an n-q plane P , the point p is an isolated point of the intersection of $V(\mathcal{I})$ and P .

This follows from the local parameterization theorem [W]. We have the

Definition 11. Let $\mathcal{I}$ be a proper ideal in $\hat{\mathcal{O}}$ or $\mathcal{O}$. We define

(11.1) $D_q(\mathcal{I}) = \inf_{\ell} D(\mathcal{I}, \ell_1, \ldots, \ell_{q-1})$,

where the infimum is taken over all choices of q-1 linear forms. We adjoint these to the ideal, and take the multiplicity.

Lemma 12. $D_q(\mathcal{I}) < \infty \Longleftrightarrow \dim V(\mathcal{I}) < q$.

Proof. This is immediate from the remark before definition 11.

We now return to our real hypersurface M . Choose a defining function, and define from it ideals $\mathcal{I}(U,p)$. We will sketch a proof below that the result will be independent of the choice of defining function.

Definition 13. $B_q(M,p) = 2 \sup_U D_q(\mathcal{I}(U,p))$.

Theorem 14. $\Delta_q(M,p) \leq B_q(M,p) \leq 2(\Delta_q(M,p))^{n-q}$. In particular, the two numbers are simultaneously finite.

Proof. The proof is similar to the proof of Theorem I.10. In case q equals one this is proved on page 630 of [D 1]. The point is that by using the ideals $\mathcal{I}(U,p)$, one estimates that $\Delta_q(M,p) \leq 2 \sup_U \tau^*(\mathcal{I}(U,p))$, where τ^* is the invariant de- defined by pulling back to one-dimensional varieties as in Proposition I.7. This invariant is smaller than or equal to the (smallest) integer s which works in 8.3, which in turn is smaller than or equal to $D(\mathcal{I}(U,p))$. (One lists the monomials in $\mathcal{O}/\mathcal{I}(U,p))$. The second inequality comes from estimating $D(\mathcal{I}(U,p))$ in terms of $\tau^*(\mathcal{I}(U,p))$ as in [D 1]. This is a sketch of the proof when q equals one. However, to prove this for higher q is easy because the q-multiplicity is defined by adjoining q-1 linear forms to the ideal (Definition 11.1). This amounts to considering the case when q equals one in a lower dimensional space.

Theorem 15. $B_q(M,p)$ is upper semi-continuous as a function of p .

Proof. Given p_0 , we must show that if p is close to p_0 , then

(15.1) $B_q(M,p) \leq B_q(M,p_0)$.

In case the right side is infinite there is nothing to prove. Suppose that it is finite. We have

(15.2) $B_q(M,p) = 2 \sup_U \inf_{\ell} D(\mathcal{I}(U,p), \ell_1, \ldots, \ell_{q-1})$.

Now, it follows from formulas 3.1, 3.2, 3.3, 4.1 and 4.2 that the coefficients of the generators of $\mathcal{I}(U,p)$ are continuous functions of U and p . By remark 9, D is usc. as a function of these parameters. Since the infimum of a family of usc. functions is usc., we have $D_q(\mathcal{I}(U,p))$ is usc.. This function is finite for p close to p_0 because of theorem 14 and theorem I.10. (The proof of the latter fact uses semi-continuity when q equals one. See [D 1].) Since it is integer valued, it has only finitely many values, so taking the supremum also preserves the upper semi-continuity. Hence 15.1 holds.

<u>Example 16</u>. Put $r(z,\bar{z}) = 2\mathrm{Re}(z_3) + |z_1^2 - z_2 z_3|^2 + |z_2|^4$. p = origin. Then $B_1(M,p) = 2D(z_3, z_1^2 - z_2 z_3, z_2^2) = 2D(z_3, z_1^2, z_2^2) = 8$. According to theorem 15, $B_1(M,p) \leq 8$ for all p close to the origin. This is to be contrasted with the fact that $\Delta_1(M,\text{origin}) = 4$, but that $\Delta_1(M,p) = 8$ for certain p arbitrarily close to the origin [D 2].

<u>Lemma 17</u>. $B_q(M,p)$ does not depend on the choice of defining equation.

<u>Proof</u>. We sketch the proof. Let r and r' be defining functions. It is enough to show that

$$(17.1) \quad \sup_U D_q(\mathscr{I}(U,p)) \leq \sup_U D_q(\mathscr{I}'(U,p)) ,$$

because we can then interchange the roles of r and r' to obtain equality. For simplicity we assume that q equals one. We write $r = 2\mathrm{Re}(h) + \|f\|^2 - \|g\|^2$ formally as in 4, and $r' = 2\mathrm{Re}(H) + \|F\|^2 - \|G\|^2$. We assume that $r' = ur$, where

$$(17.2) \quad u = 1 + \|a\|^2 - \|b\|^2 .$$

Note that (17.2) is not in the same form as (3.4), because there is no need to isolate the pure terms. This is because the ideals (h) and (H) in $\hat{\mathcal{O}}$ are equal [D 4]. So we work modulo this ideal. Given any U , we choose a copy of U for each function in a , a copy of U^* for each function in b , and consider the unitary matrix

$$(17.3) \quad U' = \begin{pmatrix} U & 0 & 0 \\ 0 & U & 0 \\ 0 & 0 & -U^* \end{pmatrix} .$$

Using (17.2) we obtain that, with evident notation,

$$(17.4) \quad F = f \oplus af \oplus bg$$

$$G = g \oplus ag \oplus bf .$$

Computing $F - U'G$ by using (17.4) and (17.3), we obtain

$$(17.5) \quad F - U'G = (f - Ug , a(f - Ug) , bU^*(f - Ug)) .$$

Hence the ideals $F - U'G$ and $f - Ug$ are the same. Therefore we must conclude that

$$(17.6) \quad \sup_U D(f - Ug , h) \leq \sup_U D(F - UG, H) .$$

This is (17.1) in case q equals one.

Remark 18. It follows from (8.3) that $B_q(M,p)$ is finitely determined, so that this invariant satisfies all the properties of Section I. It would be interesting to give an intrinsic formulation of this number.

Example 19. $r(z) = 2\mathrm{Re}(z_4) + |z_1^2 - z_2z_3|^2 + |z_2^3 - z_3^5|^2 + |z_3^7|^2$.

$B_1(M,\text{origin}) = 2D(z_4, z_1^2 - z_2z_3, z_2^3 - z_3^5, z_3^7) = 84$.

$B_2(M,\text{origin}) = 2D(z_4, z_1^2 - z_2z_3, z_2^3 - z_3^5, z_3^7, z_3) = 12$.

$B_3(M,\text{origin}) = 2D(z_4, z_1^2 - z_2z_3, z_2^3 - z_3^5, z_3^7, z_3, z_2) = 4$.

$\Delta_1(M,\text{origin}) = 14$; given by $(t^4, t^5, t^3, 0)$.

$\Delta_2(M,\text{origin}) = 6$; given by $(st, s^2, t^2, 0)$.

$\Delta_3(M,\text{origin}) = 4$; given by $(s,t,u,0)$.

III. Iterated commutators and related conditions

Kohn's [K 1] original definition of point of finite type in $\mathbb{C}^2$ involves iterated commutators. Because of the importance of this technique in partial differential equations, the notion has many applications. We recall the generalization, due to Bloom-Graham [BG], of Kohn's original definition. We also discuss related conditions on vector fields due to Bloom [B 1].

Since M is a real hypersurface of $\mathbb{C}^n$, it inherits the structure of a CR manifold. Let $TM \otimes \mathbb{C}$ denote the complexified tangent bundle to M , and let $T^{10}M$ denote the integrable subbundle of $TM \otimes \mathbb{C}$ whose sections are (1,0) vector fields tangent to M . We write, as usual, $T^{01}M$ for the complex conjugate bundle. The Levi form λ measures the extent to which $T^{10}M \oplus T^{01}M$ fails to be integrable. To define it, choose a nonvanishing real 1-form η annihilating this direct sum, and let L and L' be local sections of $T^{10}M$. We put

(1) $\lambda(L,L') = \langle d\eta, L \wedge \overline{L'} \rangle = \langle \eta, [L, \overline{L'}] \rangle$.

We now define the type of a vector field at a point.

Definition 2. Let L be a local section of $T^{10}M$, with $L(p) \neq 0$. We define the type of L at p , written type (L,p) , to be the smallest integer m so that there is an iterated commutator X of L and $\overline{L}$ of length $(m-1)$ so that

(2.1) $\langle \eta, X \rangle (p) \neq 0$.

By a commutator of length $(m-1)$, we mean a vector field

(2.2) $X = [\cdots [[A_1, A_2], A_3] \cdots A_m]$

where each A_j is either L or $\overline{L}$.

One observes that the type (L,p) equals two precisely when the Levi form, $\lambda(L,L)(p)$, does not vanish at p . More generally, we can define the types of subbundles. Let $\mathcal{B}$ be a subbundle of $T^{10}M$. We let $\mathcal{C}_m(\mathcal{B})$ denote the module over the smooth functions generated by commutators of local sections of $\mathcal{B}$ and their complex conjugates of length $m-1$.

Definition 3. The type of the subbundle $\mathcal{B}$ at p, written type$(\mathcal{B},p)$, is the smallest integer m for which there is an element X of $\mathcal{C}_m(\mathcal{B})$ for which

(3.1) $\langle \eta, X \rangle (p) \neq 0$.

Following Bloom, we can now give another family of notions of finite type.

Definition 4 [B 1]. $t_q(M,p) = \sup\{\text{type}(\mathcal{B},p)\}$, where the supremum is taken over all subbundles of $T^{10}M$ of dimension q.

This concept has no relationship to the intersection theory of M with complex analytic varieties or manifolds in general, although it seems to when M is pseudo-convex. Not much is known about the numbers $t_q(M,p)$ except when q equals $n-1$. In that case we have

Theorem 5. Let M be a smooth real hypersurface of $\mathbb{C}^n$. All of the following numbers are equal:

(5.1) $t_{n-1}(M,p)$

(5.2) $\Delta^{reg}_{n-1}(M,p)$

(5.3) $\Delta_{n-1}(M,p)$

(5.4) $\mathcal{B}_{n-1}(M,p)$

(5.5) $\mathcal{C}_{n-1}(M,p)$ (see definition III.7)

(5.6) $m_2(p)$ (see definition IV.7 and theorem IV.10.2)

Remarks on the proof. That (5.1) equals (5.2) is proved in [BG]. That (5.2) and (5.3) are equal is proved in [D 1] and [K 2]. That (5.5) equals (5.1) is proved in [B 1]. To include (5.4), note that the process of computing B_{n-1} amounts to computing the order of vanishing of the defining equation $2\,\text{Re}(z_n)$ + higher terms, where restricted to $z_n = 0$.

Unfortunately, the condition that $t_1(M,p)$ be finite has a serious disadvantage. It is not an open condition, and if assumed locally, it is not finitely determined.

Example 6. Put $r(z) = 2\,\text{Re}(z_3) + |z_1^2 - z_2^3|^2 + f(z_1,z_2)$; where p is the origin.

Suppose first that f vanishes identically. Then it is elementary to verify that type (L,p) equals 6 or 4, depending only on whether the coefficient of $\partial/\partial z_1$ in L vanishes or not. We have $t_2(M,p) = 4$, $t_1(M,p) = 6$. On the other hand, along the variety V defined by $0 = z_3 = z_1^2 - z_2^3$, we have type $(L,\tilde{p}) = \infty$, for L tangent to V and $\tilde{p}$ in V, not the origin. Thus the set where $t_1(M,p)$ is finite is not an open set. Worse yet, suppose now that f vanishes to infinite order at p, but is strongly plurisubharmonic as a function of its two variables elsewhere. Then we have type $(L,\tilde{p})$ equals 2, for all L, unless $\tilde{p}=p$, whereas type (L,p) equals 6 or 4 as before. Thus type $(L,p)<\infty$ for all L and p near the origin, so the condition (assumed locally) is not finitely determined.

There is a related invariant introduced by Bloom [B 1] and used also by Talhoui [T]. Suppose first that $\mathscr{B}$ is a subbundle of $T^{10}M$, and that trace $(\lambda(\mathscr{B}))$ denotes the trace of the restriction of the Levi form to $\mathscr{B}$. Let $C(\mathscr{B},p)$ denote the smallest integer m for which we can find $m-2$ vector fields, $L_1,\ldots,L_{m-2}$, that are local sections of $\mathscr{B}$ or complex conjugates thereof, for which

(6) $\quad (L_1L_2\cdots L_{m-2})(\text{trace}\ \lambda(\mathscr{B}))(p)\neq 0$.

Definition 7. $C_q(M,p)=\sup\{C(\mathscr{B},p)\}$, where the supremum is taken over all q dimensional subbundles $\mathscr{B}$ of $T^{10}M$.

The numbers of definition 7 satisfy properties (4.1) through (4.3) of section I, but not the last two. It is also easy to verify that

(7.1) $\Delta_q^{reg}(M,p)\leq C_q(M,p)$.

To illuminate the definition, suppose that $\mathscr{B}$ is the bundle generated by one vector field L. Then we have trace $\lambda(\mathscr{B})=\lambda(L,L)$. Thus, instead of taking iterated commutators of L and $\bar{L}$, we differentiate the Levi form applied to L and $\bar{L}$. The two notions are not identical. Let T denote a vector field transverse to $T^{10}\oplus T^{01}$ for which $\langle\eta,T\rangle=1$. Then we have, with evident notation,

(8.1) $[[L,\bar{L}],L]=[\lambda(L,L)T+A_{10}-\bar{A}_{10},L]$.

The contraction of (8.1) with η gives

(8.2) $-L\lambda(L,L)+\alpha_L\cdot\lambda(L,L)+\lambda(L,A_{10})$

for some function α_L, not in general zero. Thus, even up to order three, there are terms in the iterated commutator that do not appear in a derivative of the Levi form. Note, however, that if λ is assumed semi-definite, then the term $\lambda(L,A_{10})$ must vanish if $\lambda(L,L)$ does. In that case one expects that the two invariants, type (L,p) and $C(\mathscr{B},p)$, ($\mathscr{B}$ = bundle generated by L) are equal. Bloom [B 1] conjectures even more; namely that, if λ is semi-definite, then

Conjecture 8. (M pseudoconvex) $C_q(M,p)=t_q(M,p)=\Delta_q^{reg}(M,p)$.

We now turn to an interesting example in the case where M is not pseudoconvex.

Example 9 [B 1]. Put $r(z)=2\,\mathrm{Re}(z_3)+|z_1|^4+2|z_1|^2+2(z_2+\bar{z}_2)|z_1|^2$; let p be the origin. Let

$$L=\frac{\partial}{\partial z_1}-\bar{z}_1\frac{\partial}{\partial z_2}+g\frac{\partial}{\partial z_3}\quad,$$

where g is chosen so L will be tangent to M. It is elementary to verify that

$[[L,\bar{L}],L]$ vanishes identically. We have

(9.1) $t_1(M,p) = C_1(M,p) = \infty$; $t_2(M,p) = C_2(M,p) = 2$

(9.2) $\Delta_1^{reg}(M,p) = \Delta_1(M,p) = 4$; $\Delta_2^{reg}(M,p) = \Delta_2(M,p) = 2$.

From this example one sees that these numbers C_q and t_q are not related to orders of contact in the general case. One does always have that

(9.3) $\Delta_q^{reg}(M,p) \leq t_q(M,p)$.

Bloom [B 1] has proved conjecture 8 in the special case where M lies in $\mathbb{C}^3$. Talhoui [T] proves a weak version of it in $\mathbb{C}^n$. Because of his assumptions on the rank of the Levi form, this result is not really any more general. The author [D5] has proved a related result. Let M be any CR manifold, and suppose that the Levi form is semi-definite on the span of the (1,0) parts of all of the iterated commutators of a (1,0) vector field L and its complex conjugate. Let c denote the number $C(\mathcal{B},p)$, where $\mathcal{B}$ is the bundle generated by L , and let N denote type (L,p) . Then $c \leq \max(N, 2N-6)$. Perhaps c and N are equal under these hypotheses.

One sees from example 6 that the invariants t_q and C_q are not locally finitely determined. This casts doubt upon their usefulness; however they have the advantage that they can be defined on any CR manifold, and if Bloom's conjecture were true, one could formulate intrinsically the number $\Delta_q^{reg}(M,p)$. Since none of these numbers arise in subelliptic estimates, the interest in them has subsided somewhat.

On the other hand, iterated commutators play a large role in the theory of scalar partial differential operators. The interested reader should consult Hörmander [H] and the references there. We mention one well-known example:

<u>Theorem 10</u> (Hörmander [H]; see also Kohn [K3]). Suppose that $L_0, L_1, \ldots, L_N$ are vector fields defined near 0 in $\mathbb{R}^n$. Let P denote the second order operator

$$P = L_0 + \sum_1^N L_j^2 \quad .$$

Then, P is subelliptic $\Longleftrightarrow$ there is an integer m so that the L's and their commutators of length less than or equal to m span the tangent space at 0 in $\mathbb{R}^n$.

The $\bar{\partial}$-Neumann problem on $(0,n-1)$ forms exhibit essentially the same properties as in Theorem 10. In this case the iterated commutators control the situation [K2]. However, the finite type conditions of sections I and II are required in the $\bar{\partial}$-Neumann problem on $(0,q)$ forms. Presumably variations of those ideas will find their way into the (as yet not understood) study of subelliptic systems.

IV. <u>Subelliptic estimates and Catlin's multi-type</u>

We will not describe the $\bar{\partial}$-Neumann problem in any detail here. See [K2,C1]. However, if Ω is a bounded pseudoconvex domain with smooth boundary, and α is a $\bar{\partial}$-closed $(0,q)$ form with L^2 coefficients, then the $\bar{\partial}$-Neumann problem constructs a particular solution to the equation $\bar{\partial}u = \alpha$. One obviously wishes to know when u must be smooth wherever α is smooth. This local regularity property follows from the subelliptic estimates described below.

Suppose that p is a point in $bd\Omega$. The $\bar{\partial}$-Neumann problem satisfies a subelliptic estimate on $(0,q)$ forms at p if there are positive constants C, ε and a neighborhood U of p such that

$$(1) \qquad \|\phi\|_{\varepsilon}^{2} \leq C(\|\bar{\partial}\phi\|^{2} + \|\bar{\partial}^{*}\phi\|^{2} + \|\phi\|^{2})$$

whenever ϕ is a smooth form, in the domain of $\bar{\partial}^{*}$, and supported in U. It follows from the work of Kohn that ε can be no larger than $1/2$; this occurs when $bd\Omega$ is strongly pseudoconvex at p. The search for geometric conditions equivalent to the estimate for some ε smaller than $1/2$ has motivated most of the work described in this paper.

Kohn [K2] invented an iterative process for proving the estimate 1; his method gives only a rough approximation to the actual value of ε. He considered the ideal of germs of smooth functions f at p such that there is some positive ε for which 1 holds, when the left side is replaced by $\|f\phi\|_{\varepsilon}^{2}$. One then wants to find conditions that imply that the constant function 1 lies in the ideal. There are certain functions, for example appropriate minor determinants of the Levi form, that lie in this ideal. By the processes of differentiating, evaluating determinants and computing real radicals of an ideal, Kohn defines ideals $I_k^q(p)$ iteratively. He proves that

2. If 1 lies in $I_k^q(p)$ for some k, then there is a subelliptic estimate on $(0,q)$ forms.

3. If $M = bd\Omega$ is a real analytic hypersurface, then 1 lies in $I_k^q(p) \Longleftrightarrow M$ contains no germ of a complex analytic q dimensional variety at p. This relies on the Diederich-Fornaess theorem [DF]. In particular, if M is the boundary of a bounded domain, and is real analytic, this must be true.

Combining this with the results of Catlin below, we have the following implications:

4. C^∞ case

$$1 \text{ in } I_k^q(p) \Longrightarrow \text{subelliptic estimate} \Longleftrightarrow \Delta_q(M,p) < \infty .$$

5. Real analytic case

$$1 \text{ in } I_k^q \Longleftrightarrow \text{subelliptic estimates} \Longleftrightarrow \Delta_q(M,p) < \infty .$$

$$\Longleftrightarrow M \text{ contains no germ of a } q \text{ dimensional complex analytic variety at } p .$$

It remains important to verify the remaining implication in 4, because Kohn's ideals can be defined on CR manifolds and have other applications as well [K4].

Catlin's proof of the estimates differs in technical respects from Kohn, as he constructs plurisubharmonic function with large Hessians. In his proof he also uses the concept of a multi-type. He associates an n-tuple of positive rational numbers (or infinity) to a boundary point. His numbers are ordered oppositely from the properties 1.1 through 1.5 of section I. They also involve some of the concepts of section III. Here we do little more than recall the definition, list the basic properties, and give some examples. We do warn the reader that a finite multi-type does not guarantee that Δ_1 be finite; however, in case Δ_1 is finite, the multi-type also is and its properties are basic to his proof of the estimates.

<u>Definition 6</u>. A weight is an n-tuple of numbers λ_i, with $1 \le \lambda_i \le \infty$, and $\lambda_i \le \lambda_{i+1}$ for all i. These weights are ordered lexicographically. We also demand that $1 = \lambda_1$ and that λ_k is defined, if it is finite by

(6.1) $\sum^{k} a_j \lambda_j^{-1} = 1 \qquad a_k > 0\ , \quad a_j \ge 0$ are integers.

<u>Definition 7</u>. Suppose that M is a real hypersurface of $\mathbb{C}^n$, with defining function r. A weight $(\lambda_1, \lambda_2, \ldots, \lambda_n)$ is distinguished at p if, in some coordinate system

(7.1) $\sum^{n} (\alpha_i + \beta_i)\lambda_i^{-1} < 1 \Longrightarrow D^{\alpha}\bar{D}^{\beta} r(p) = 0$.

The multi-type $\mathcal{M}(p)$ is the smallest weight (in the lexicographical ordering) that is larger than all distinguished weights.

Thus, if $\mathcal{M}(p) = (m_1, m_2, \ldots, m_n)$ we can think of m_i as a weight to associate with the coordinate direction z_i. However, one inductively uses the previous weights to define them.

<u>Examples 8</u>.

(8.1) $r(z) = 2\,\mathrm{Re}(z_n) + \sum_{1}^{n-1} |z_j|^{2p_j}, \quad p_1 \le p_2 \le \cdots \le p_{n-1}$

$\mathcal{M}(0) = (1, 2p_1, 2p_2, \ldots, 2p_{n-1})$

(8.2) $r(z) = 2\,\mathrm{Re}(z_3) + |z_1^2 - z_2^3|^2$

$\mathcal{M}(0) = (1,4,6)$

(8.3) $r(z) = 2\,\mathrm{Re}(z_3) + |z_1|^4 + |z_2|^8 + 2\,\mathrm{Re}(\bar{z}_1 z_2^5)$

$\mathcal{M}(0) = (1,4,20/3)$

$\Delta_2^{\mathrm{reg}}(M,0) = 4 \ ; \ \Delta_2^{\mathrm{reg}}(M,0) = 8$

<u>Observation 9</u>. Consider example (8.2). Let p be a point along the variety $z_1^2 = z_2^3$ close to 0 and lying in M. Then the multi-type at p will be $(1,2,\infty)$, unless p is the origin. Thus there are points close by a given point

of finite multi-type, with infinite multi-type. However, in the lexicographical sense, the weight $(1,2,\infty)$ is smaller than $(1,4,6)$. This sense of upper semi-continuity holds and is one of the main points in Catlin's work.

Theorem 10 (Catlin). Let M be pseudoconvex. Then the multi-type $\mathcal{M}(z)$ satisfies the following properties:

10.1 $\mathcal{M}$ is upper semi-continuous in the lexicographical sense; i.e., if p_0 in M is given, then there is a nbd. U of p_0 on which

$$\mathcal{M}(p) \leq \mathcal{M}(p_0)$$

10.2 If $\mathcal{M}(p_0) = (m_1,\dots,m_n)$, then $m_{n+1-q} \leq \Delta_q(M,p_0)$. If $q = n-1$, we have $m_2 = \Delta_{n-1}(M,p_0)$.

10.3 Suppose m_{n-q} is finite. Then the set of points p close to p_0 for which $\mathcal{M}(p) = \mathcal{M}(p_0)$ is contained in a submanifold of holomorphic dimension at most q. See [C2] or [K1] or [DF] for the concept of holomorphic dimension. Thus 10.3 gives us a stratification of M.

10.4 Suppose $\mathcal{M}(p)$ is finite. Then after a change of coordinates, $\mathcal{M}(p)$ is distinguished in the sense of 7.1.

The proof of this theorem is quite long and involves an alternative definition which Catlin calls the commutator multi-type. See [C2]. The key property for subellipticity is the stratification given by 10.3. It follows from theorem I.10 of this paper and property 10.3 that M is a finite union of sets which are locally contained in manifolds of holomorphic dimension q-1. Thus, except for points on a set that is negligible in this sense, the multi-type is smaller nearby, hence "better behaved".

The reader should consult [C3] when it appears, where even more precise versions of these results are required to establish the subelliptic estimates. Also, Catlin has an alternative measurement of the order of contact of q-dimensional varieties.

V. Essential finiteness

In this section we discuss an important concept due to Baouendi-Jacobowitz-Treves [BJT]. They proved the following important

Theorem 1. Suppose that $\tilde{M}$ and M are real analytic hypersurfaces of $\mathbb{C}^n$ and that $\Phi : \tilde{M} \to M$ is a CR diffeomorphism. Then, if M is essentially finite, Φ is actually real analytic.

We give the definition of essential finiteness and several propositions that show its relationship to the other notions. We suppose that M is given by the vanishing of a real analytic function $r(z,\overline{z})$, that 0 lies in M, and we give the definition that M be essentially finite at 0. M is then essentially finite if it is essentially finite at each point.

Definition 2. M is essentially finite at 0 if the equations $r(0,\zeta) = 0$ and $r(z,\zeta) = 0$ imply that $z = 0$.

In fact this definition is independent of the choice of defining function. The following propositions give some insight into this concept.

Proposition 3. If M is essentially finite at p, then $\Delta_{n-1}(M,p)$ is finite. If $\Delta_1(M,p)$ is finite, then M is essentially finite at p. Unless $n = 2$, the concept is not equivalent to any of the intermediate notions considered earlier.

Proof. See [BJT]; both are easy. Note that neither converse is true unless n equals 2.

We have seen in II.3.4 that every real analytic function can be written as

(4) $$2\,\mathrm{Re}(h) + \|f\|^2 - \|g\|^2 .$$

In case g vanishes, we have equality between 1-finite-type and essential finiteness.

Proposition 5. Suppose M is defined by $2\,\mathrm{Re}(h) + \|f\|^2$. Then M is essentially finite at $0 \iff \Delta_1(M,0) < \infty$.

Proof. $\Longleftarrow$ this is part of proposition 3.
$\Longrightarrow$ To prove that $\Delta_1(M,0)$ is finite, it is enough to show that $V(h,f) = \{0\}$. See theorem II.6. Suppose that $z \in V(h,f)$. Then $r(z,\zeta) = \overline{h}(\zeta)$. On the other hand, the definition of essential finiteness tells us that, if $r(z,\zeta) = 0$ whenever $r(0,\zeta) = \overline{h}(\zeta) = 0$, then z equals 0. Thus $V(h,f)$ is trivial and we are finished.

Finally we give a necessary and sufficient condition that is easy to verify. Let M be defined by r, which we write, after a local coordinate change, as

(6) $$r(z,\bar{z}) = 2\mathrm{Re}(z_n) + \Sigma h_b(z)\bar{z}^b = 2\mathrm{Re}(z_n) + \Sigma h_a(z)(\bar{z}')^a + \bar{z}_n S .$$

Here each h_b vanishes at 0, each multi-index a satisfies $a_n=0$, z' denotes the first n-1 coordinates of z, and the sum S is determined. Let V denote the (germ of a) variety defined by z_n and the h_a, where $a_n=0$.

Proposition 7. M is essentially finite at 0 if and only if V is trivial.
Proof. Note that the condition that $r(0,\zeta)$ vanishes amounts to $\zeta_n=0$. For such ζ, the condition that $r(z,\zeta)$ vanishes becomes

(8) $$0 = z_n + \Sigma h_a(z)(\zeta')^a .$$

If V is not trivial, choose a non-zero w in V. Equation 8 then holds for z=w, so M is not essentially finite. Conversely, if V is trivial, and 8 holds, then each coefficient in the power series vanishes, so z lies in V. Since V is trivial, z vanishes, and M is essentially finite.

VI. Open questions

1. Prove the implication of IV.4 that $\Delta_q(M,p)$ is finite implies that 1 lies in $I_k^q(p)$.

2. Give an intrinsic treatment of "orders of contact" and "multiplicities" on CR manifolds.

3. Prove Bloom's conjecture III.8.

4. Determine the boundary behavior of the Bergman kernel, and its relationship to these invariants. See [He] and [DFH].

5. What are the important invariants for subellipticity in case M is not pseudoconvex? See [Ho].

6. Determine approach regions for the boundary behavior of holomorphic functions on weakly pseudoconvex domains. See [NSW] for the situation when $\Delta_{n-1}(M,p)$ is finite.

7. Find normal forms for the defining equations of real analytic hypersurfaces with degenerate Levi form. See [D4].

Bibliography

[BJT] M. S. Baouendi, H. Jacobowitz and F. Treves, On the analyticity of CR mappings, Annals of Math. 122 (1985) 365-400.

[Be 1] S. Bell, Biholomorphic mappings and the $\bar\partial$-problem, Annals of Math. 14 (1981) 103-113.

[B1] T. Bloom, On the contact between complex manifolds and real hypersurfaces in $\mathbb{C}^3$, Trans. A.M.S., Vol. 263, No. 2 (1981) 515-529.

[B2] ________, Remarks on type conditions for real hypersurfaces in $\mathbb{C}^n$, pp. 14-24 in Several Complex Variables, Proc. of Internat. Conf. at Cortona, Italy, 1978.

[BG] T. Bloom and I. Graham, A geometric characterization of points of type m on real hypersurfaces, J. Diff. Geometry 12 (1977) 171-182.

[C1] D. Catlin, Necessary conditions for subellipticity of the $\bar\partial$-Neumann problem, Annals of Math. 117 (1983) 147-171.

[C2] ________, Boundary invariants of pseudoconvex domains, Annals of Math. 120 (1984) 529-586.

[C3] ________, Subelliptic estimates for the $\bar\partial$-Neumann problem on pseudoconvex domains, preprint.

[D1] J. D'Angelo, Real hypersurfaces, orders of contact, and applications, Annals of Math. 115 (1982) 615-637.

[D2] ________, Subelliptic estimates and failure of semi-continuity for orders of contact, Duke Math. J. 47 (1980) 955-957.

[D3] ________, Intersection theory and the $\bar\partial$-Neumann problem, Proc. of Symposia in Pure Mathematics (1984), Vol. 41, 51-58.

[D4] ________, Defining equations for real analytic hypersurfaces, Trans. A.M.S. Vol. 295, No. 1, (1986) 71-84.

[D5] ________, Iterated commutators and derivatives of the Levi form, preprint.

[DF] K. Diederich and J. Fornaess, Pseudoconvex domains with real analytic boundaries, Annals of Math. 107 (1978) 371-384.

[DFH] K. Diederich, J. Fornaess and G. Herbort, Boundary behavior of the Bergman metric, Proc. of Symposia in Pure Mathematics (1984), Vol. 41, 59-67.

[G] P. Greiner, On subelliptic estimates of the $\bar{\partial}$-Neumann problem in $\mathbb{C}^2$, J. Diff. Geometry 9 (1974) 239-250.

[He] G. Herbort, The boundary behavior of the Bergman kernel function and metric for a class of weakly pseudoconvex domains of $\mathbb{C}^n$, Math. Z. 184 (1983) 193-202.

[Ho] L. Ho, Subellipticity of the $\bar{\partial}$-Neumann problem on non-pseudoconvex domains, thesis, Princeton University, 1983.

[H] L. Hörmander, The Analysis of Linear Partial Differential Operators III, IV, Springer-Verlag, 1985.

[K1] J. Kohn, Boundary behavior of $\bar{\partial}$ on weakly pseudoconvex manifolds of dimension two, J. Diff. Geometry 6 (1972) 523-542.

[K2] ________, Subellipticity of the $\bar{\partial}$-Neumann problem on pseudoconvex domains: sufficient conditions, Acta Math., Vol. 142 (1979) 79-122.

[K3] ________, Pseudodifferential operators and hypoellipticity, Proc. of Symposia in Pure Mathematics (1973), Vol. 23, 61-69.

[K4] ________, Estimates for $\bar{\partial}_b$ on pseudoconvex manifolds, Proc. of Symposia in Pure Mathematics (1985), Vol. 43, 207-217.

[NSW] A. Nagel, E. Stein and S. Wainger, Boundary behavior of functions holomorphic in domains of finite type, Proc. Nat. Acad. of Sci. 78 (1981), No. 11, 6596-6599.

[S] I. Shafarevich, Basic Algebraic Geometry, Springer-Verlag, Berlin and New York, 1977.

[T] A. Talhoui, Conditions suffisantes de sous-ellipticité pour $\bar{\partial}$, C. R. Acad. Sci. Paris, t. 296 (1983), 427-429.

[W] H. Whitney, Complex Analytic Varieties, Addison Wesley Publishing Co., 1972.

Iterated Commutators and Derivatives of the Levi Form

John P. D'Angelo*
University of Illinois
Urbana, Illinois 61801

Introduction

The theory of weakly pseudoconvex domains in $\mathbb{C}^n$ often involves finite type conditions [1,2,3,4,5,6,7]. There are many ways to measure the degeneracy of the Levi form; all involve taking higher derivatives. Some of the recently popular notions come from algebraic geometry, such as the intersection theory of complex analytic varieties with the boundary [5]. In this paper, however, we will be concerned with differential geometric notions, such as iterated commutators. This notion first arose in Kohn's study of $\bar{\partial}$ [7]. Our setting will be CR manifolds; these are real manifolds whose tangent spaces have a certain amount of complex structure. We consider two type conditions for a (1,0) vector field L at a point p. The first, type (L,p), measures the number of commutators of L and its complex conjugate that are required to obtain a component in the so called "bad direction", the tangential direction that is not part of the holomorphic tangent space [1,7]. The second, written $C(L,p)$, and first defined by Bloom [1], involves differentiating the Levi form in the directions defined by L and its conjugate. See also [6]. Bloom has conjectured that various invariants formed from these notions are equal, in case M is a pseudoconvex hypersurface in $\mathbb{C}^n$.

In this paper we formulate a related conjecture for the case of a single vector field. This conjecture is the following. Suppose that M is a CR manifold, and that the Levi form is semi-definite on the span of L, and the (1,0) parts of all its iterated commutators up to sufficiently high order. Then type (L,p) equals $C(L,p)$.

For CR manifolds of dimension 3, the result is trivial and does not require a pseudoconvexity hypothesis. For pseudoconvex hypersurfaces in $\mathbb{C}^3$, the result can be derived from the work of Bloom [1]. Such hypersurfaces are CR manifolds of dimension 5. It is also easy to verify this for certain particular vector fields, such as a vector field of minimal type, but the general case is open.

In this paper we obtain some partial results. First of all, we show that the conjecture holds for vector fields of type 4. The result is trivial for vector fields of type 2, and in the pseudoconvex case, there cannot be vector fields of type 3. Secondly, we give an estimate for $C(L,p)$ in case type (L,p) is known to equal N. A simplified form of this estimate gives the result that

* Partially supported by the NSF Grants MCS-8108814(A04) and DMS-8501008 and by The Institute for Advanced Study.

$C(L,p) \leq \max(N, 2N-6)$. To prove this result, we write out formulas for iterated commutators, and identify the terms as Levi forms applied to the original vector field and (1,0) parts of its higher commutators. By differentiating appropriate minor determinants of the Levi form at points of degeneracy, we are able to obtain an estimate for $C(L,p)$. See theorem 22.

At no time in these proofs do we make any choice of coordinates. Only the CR structure is used. The formulas here could perhaps be of use in problems on weakly pseudoconvex CR manifolds, because they do not appeal to such theorems as the Diederich-Fornaess theorem [6].

I acknowledge helpful discussions with Tom Bloom on many of these ideas. In fact the motivation for this paper comes from trying to prove some of the conjectures and results in [1] without using his reduction to the homogeneous case. I also thank J. J. Kohn for originally leading me to the study of iterated commutators.

Preliminaries

A CR manifold is a real manifold whose tangent spaces have a certain amount of complex structure. To state this more precisely, let M be a real manifold, and let $TM \otimes \mathbb{C}$ denote its complexified tangent bundle. M is called a CR manifold if there is an integrable subbundle, $T^{10}M$, of $TM \otimes \mathbb{C}$ with the following property:

1. The intersection of $T^{10}M$ with its complex conjugate bundle $T^{01}M$ consists of the zero section alone.

Henceforth in this paper we will also assume that the bundle sum $T^{10}M \oplus T^{01}M$ has codimension one in $TM \otimes \mathbb{C}$. This bundle sum is called the holomorphic tangent bundle. Thus M is an abstraction of a real hypersurface in a complex manifold. In that case local sections of $T^{10}M$ are those vector fields that are combinations of the $\partial/\partial z_j$, where the z_j are local coordinates. We call local sections of $T^{10}M$, (1,0) vector fields. We denote by π_{10} and π_{01} the projections onto the respective subbundles.

To describe our results on iterated commutators, we recall the definition of the Levi form. Notice first that there is a purely imaginary non-vanishing one form η , defined up to a multiple, that annihilates the holomorphic tangent bundle.

2. Definition Let L, K be local sections of $T^{10}M$. We define the Levi form, as usual, by the formula

$$\lambda(L,\overline{K}) = \{\eta, [L,\overline{K}]\} . \tag{2.1}$$

Here $\{\ ,\ \}$ denotes contraction and $[\ ,\ ]$ denotes the commutator. By the Cartan formula, we have

$$\lambda(L,\overline{K}) = \{-d\eta, L\wedge\overline{K}\} \tag{2.2}$$

Let L be a (1,0) vector field. We write L^{*m} for any vector field that is an iterated commutator of the form

$L^{*m} = [\ldots,[[L_1,L_2],\ldots L_m]$, where each L_j is either L or $\overline{L}$.
We write $\mathscr{L}^m$ for any vector field that is of the form $\pi_{10}(L^{*m})$ or its conjugate. We will also need a notation for orders of vanishing of functions with respect to L and $\overline{L}$. Let p be a point in M , and let f be the germ of a smooth function at p . Put

$$v_{L,p}(f) = v_L(f) = j ,$$

if j is the smallest integer such that there is a monomial $\mathscr{D}^j$ in L and $\overline{L}$, of order j , for which $\mathscr{D}^j(f)(p)$ is non-zero. Finally, we also write $\mathscr{D}^j$ for any differential operator of order j that is formed from L and $\overline{L}$. In the definition of v_L , however, we only allow monomials. This is because commutators can be of lower order. The reader can exhibit easily examples where $v_L(f)$ is larger than $v_X(f)$, for $X = [L,\overline{L}]$. We state as a remark two facts about v_L .

<u>3. Remarks</u>. $v_L(fg) = v_L(f) + v_L(g)$. If, also, $0 \leq f \leq g$, then $v_L(g) \leq v_L(f)$.

We recall two type conditions on (1,0) vector fields.

<u>4. Definition</u>. Let L be a local section of $T^{10}M$, defined near p . We say that the type of L at p is m , if m is the smallest positive integer for which there is a vector field L^{*m} such that

$\langle L^{*m},\eta\rangle(p)$ is non-zero.

If no such m exists, we say that type (L,p) equals infinity.

<u>5. Definition</u>. Let L be a local section of $T^{10}M$, defined near p . We say that $C(L,p)$ equals m , if $v_{L,p}(\lambda(L,\overline{L})) = m-2$.

These type conditions are not identical; our purpose in this paper is to investigate their relationship. Obviously the numbers are simultaneously two, but either number can be three without the other number being three. Let us briefly consider that case. We can write

$$[L,\overline{L}] = A + \overline{B} + \lambda(L,\overline{L})T \text{ , where } T \text{ is purely imaginary, and } \langle T,\eta\rangle = 1 \quad (6.1)$$

Then we have

$$\langle[[L,\overline{L}],L],\eta\rangle = -\lambda(L,\overline{B}) - L(\lambda(L,\overline{L})) + \lambda(L,\overline{L})\langle[T,L],\eta\rangle \text{ .} \quad (6.2)$$

Thus if $\lambda(L,\overline{L})$ vanishes at p , we see that the first term on the right side of 6.2 enables the whole expression to vanish at p without the second term vanishing, and conversely the vanishing of the second term does not in general force the vanishing of the first. If, however, λ is assumed to be semi-definite, then 6.2 must vanish if $\lambda(L,\overline{L})$ does, so neither type can be three in that case. Here we use two elementary facts: a first derivative of a function at a critical point must vanish, and $\lambda(L,\overline{B})$ must vanish if $\lambda(L,\overline{L})$ does, when λ is semi-definite. These facts will be used repeatedly below.

7. Definition. M is called pseudoconvex at p if the form λ is semi-definite. We can always choose the sign of η so that λ is positive semi-definite.

The presence of the third term on the right of 6.2 suggests the definition of the following one form, depending on T.

8. Definition. If X is any local section of $TM \otimes \mathbb{C}$, we put

$$\{\alpha,X\} = \{[T,X-\{X,\eta\}T],\eta\} .$$

We have omitted the dependence on T from the notation. We will usually write $\{\alpha,X\}$ as α_X. Note also that α is real. This form was defined by the author in [3], but has been little used. We give an alternate description of this form as a Lie derivative, even though we do not need this result.

9. Proposition. $\alpha = -\mathcal{L}_T\eta$ = minus the Lie deriv. of η in the direction T.

Proof.
$$\begin{aligned}\{\mathcal{L}_T\eta,X\} &= \{i_T d\eta + d\, i_T\eta,X\} = \{d\eta, T\wedge X\} + X(i_T\eta) = \{d\eta,T\wedge X\}\\ &= T\{\eta,X\} - X\{\eta,T\} - \{\eta,[T,X]\}\\ &= T\{\eta,X\} - \{\eta,[T,X]\}\\ &= -\{\alpha,X\} .\end{aligned}$$

We have used the usual formula for Lie derivatives, the fact that $i_T\eta = 1$ is a constant, the Cartan formula, the fact that $\{\eta,T\}=1$ again, and finally the definition of α.

We are now ready to write down formulas for a general iterated commutator. Let L^{*m} denote an iterated commutator of the form

$$L^{*m} = [\ldots[[L_1,L_2],\ldots L_m] ,$$

where each L_i is either L or $\overline{L}$. We assume that $L_1 = L$ and $L_2 = \overline{L}$. Then we have

10. Proposition. $\{L^{*m},\eta\}$ equals

$$(\alpha_L - L)\{L^{*m-1},\eta\} - \lambda(L,\pi_{01}L^{*m-1}) \text{ or} \tag{10.1}$$

$$(\overline{\alpha_L - L})\{L^{*m-1},\eta\} + \lambda(\pi_{10}L^{*m-1},\overline{L}) . \tag{10.2}$$

We have 10.1 if the last vector field L_m is L, and 10.2 if it is $\overline{L}$.

Proof. Suppose for concreteness that L_m equals L. We can write

$$L^{*m-1} = \pi_{10}L^{*m-1} + \pi_{01}L^{*m-1} + \{L^{*m-1},\eta\}T . \tag{10.3}$$

Taking the commutator of 10.3 with L, using the integrability of $T^{10}M$, and noting the definitions of α and λ, we obtain the desired formula 10.1.

11. Corollary. Suppose that L and $\overline{L}$ alternate in the above commutator. We then have the formula that $\{L^{*2k+2},\eta\}$ equals

$$\Sigma((\overline{\alpha_L - L})(\alpha_L - L))^j\lambda(X_j,\overline{L}) + \Sigma((\overline{\alpha_L - L})(\alpha_L - L))^j(\overline{\alpha_L - L})\lambda(L,Y_j) . \tag{11.1}$$

In this formula, the first sum runs from 0 to k, the second from 0 to $k-1$, and the vector fields X_j and Y_j are defined as follows.

$$X_k = L\ ,\quad X_j = \pi_{10}[L^{*2k-2j}, L] \quad \text{for } j \text{ less than } k\ . \tag{11.2}$$

$$Y_j = -\pi_{01}L^{*2k-2j} \tag{11.3}$$

Note the minus sign in the definition of Y_j . This arises because the (1,0) vector field arises first in the definition of λ .

12. Corollary. For any iterated commutator L^{*m} , there are operators $\mathscr{D}^k$ so that $\{L^{*m},\eta\} = \Sigma\mathscr{D}^{m-2-j}(g_j)$, where the sum extends from 0 to m-2 , and $g_j = \lambda(L,\mathscr{L}^{j+1})$ or a conjugate thereof.

Proof. This formula follows by induction from proposition 10. Note that we have absorbed the α_L into the operators $\mathscr{D}^{m-2-j}$.

Results

We can now prove that, under a pseudoconvexity hypothesis, the two types are equal to four simultaneously. We present this proof in perhaps excessive detail, because it motivates the proof of theorem 22.

13. Theorem. Let M be a CR manifold. Suppose that the Levi form is semi-definite on the span of $L,\pi_{10}[L,\bar{L}]$, and $\pi_{10}[[L,\bar{L}],L]$. Then, type $(L,p) = 4$ if and only if $C(L,p) = 4$.

Proof. Suppose first that $C(L,p)$ equals four. Because of our positivity assumption, and the fact that the type is not two, the function $\lambda(L,\bar{L})$ is non-negative and vanishes at p . Letting $L = A + Bi$, we see that $\bar{L}L$ can be expressed as $A^2 + B^2 + i[A,B]$. Since $[A,B]$ is first order, we see from calculus that

$$\bar{L}L\lambda(L,\bar{L})(p) \text{ is non-negative.} \tag{14}$$

If 14 vanishes, then both A^2 and B^2 applied to $\lambda(L,\bar{L})$ vanish at p . From this, and the formula that $L^2 = A^2-B^2+i(AB+BA)$, we see that the analog of 14 with both derivatives being L must also vanish. Since $C(L,p)$ is four, this cannot be, so we must have that 14 is actually strictly positive. We now show that the contraction $\{[[[L,\bar{L}],L],\bar{L}],\eta\}(p)$ equals the sum of 14 and a non-negative term. Thus the type of L will be four, since it is assumed not two, and we have seen that it is not three, just before definition 7. To verify the value of the contraction, we use formula 10.2. This becomes, after ignoring those terms that obviously vanish at p ,

$$\begin{aligned}\{L^{*4},\eta\}(p) &= \overline{(\alpha_L-L)}(\alpha_L-L)\lambda(L,\bar{L}) + \overline{(\alpha_L-L)}\lambda(L,-\pi_{01}[L,\bar{L}]) + \lambda(\pi_{10}[[L,\bar{L}],L],\bar{L}) \\ &= (\bar{L}L\lambda(L,\bar{L}) + \bar{L}\lambda(L,\pi_{01}[L,\bar{L}]))(p)\end{aligned} \tag{15}$$

We must show that the second term in 15 is non-negative. To do so, we write $\bar{X}$ for $\pi_{01}[L,\bar{L}]$ and use the Jacobi identity. This gives

$$\{[[L,\bar{X}],\bar{L}]\eta\} = -\{[[\bar{X},\bar{L}],L],\eta\} - \{[[L,\bar{L}],\bar{X}],\eta\}$$
$$= \lambda(L,[\bar{X},\bar{L}]) - \lambda(X,\bar{X}) \tag{16}$$

The first term on the right side of 16 vanishes, because λ is semi-definite on this space. The second term is of course non-positive. We compare with the left hand side. It becomes

$$-\bar{L}\lambda(L,\bar{X}) + \text{terms that vanish at } p . \tag{17}$$

We obtain from 16 and 17 that $\bar{L}\lambda(L,\bar{X})$ must be non-negative at p. This completes the first half of the proof.

Now we suppose that type (L,p) equals four. We have the right side of 15, or else the analogous term arising from a commutator of the form $Z = [[[L,\bar{L}],L],L]$. First we assume the the right side of 15 is non-zero at p ; below we show that the possibility arising from the analogous commutator is not actually possible. So, if 15 is non-zero at p, we claim that the first term on the right of 15 is itself positive. Consider the 2 by 2 determinant

$$f = \lambda(L,\bar{L})\lambda(X,\bar{X}) - |\lambda(L,\bar{X})|^2 \tag{18}$$

Because of the non-negativity of λ , f is non-negative. Also f vanishes at p , so by calculus (and the remark before 14) we have

$$\bar{L}Lf \text{ is non-negative at } p . \tag{19}$$

This gives us, if we also assume that $\bar{L}L\lambda(L,\bar{L})$ vanishes at p , that

$$|\bar{L}\lambda(L,\bar{X})|^2 + |L\lambda(L,\bar{X})|^2 \text{ vanishes at } p . \tag{20}$$

Putting this all together tells us that $\bar{L}L\lambda(L,\bar{L})(p)$ must be positive. The remaining possibility from the commutator Z cannot occur. For, if $\bar{L}L\lambda(L,\bar{L})$ vanishes, then from 20 we obtain that $L\lambda(L,\bar{X})$ vanishes. Thus the contraction $\{Z,\eta\}$ equals $L^2\lambda(L,\bar{L})$. This is the first term in the analog of 15. Since we have seen earlier that this must vanish when $\bar{L}L\lambda(L,\bar{L})$ does, this contraction can be non-zero only when 15 is non-zero. Thus we have proved that if type (L,p) is four, then $\bar{L}L\lambda(L,\bar{L})$ does not vanish at p , so $C(L,p)$ equals four also. Note that similar arguments show that, if M is pseudoconvex, then $C(L,p)$ is always even if it is finite.

We now turn to the general case. First we recall the formula in Corollary 12.

<u>21. Formula</u> $\{L^{*N},\eta\} = \Sigma\mathscr{D}^{N-2-j}(g_j)$, where $\mathscr{D}^j$ are operators of order j formed from L and $\bar{L}$, and g_j are functions that can be written in the form

$$g_j = \lambda(L,\bar{\mathscr{L}}^{j+1}) \text{ or a conjugate thereof.} \tag{21.1}$$

<u>Remark.</u> Note that the total number of derivatives in each term equals N ; j on the operator, 1 on L or $\bar{L}$, and the rest coming from an earlier iterated commutator.

We now prove a theorem that gives a bound for $C(L,p)$ when we know that type (L,p) is finite. If type (L,p) equals N, the result tells us in particular that $c(L,p) \leq 2N-2$.

__22. Theorem.__ Let L be a $(1,0)$ vector field on a CR manifold M. Suppose that the Levi form λ is semi-definite on the span of L and all the $(1,0)$ parts of its iterated commutators up to order N. Suppose that type (L,p) equals N. Let L^{*N} be an iterated commutator such that $\{L^{*N},\eta\}(p)$ does not vanish. Let $\mathcal{L}^j$, for $j=1$ to $N-2$, denote the $(1,0)$ parts of the iterated commutators up to order j coming from L^{*N}. Finally, let v_j denote the number

$$v_L(\lambda(\mathcal{L}^{j+1},\overline{\mathcal{L}}^{j+1})).$$

Then, $C(L,p) \leq \max((2N-2-2j-v_j),N) \leq 2N-2$. (The max is taken over all j). In fact, $C(L,p) \leq \max(N,2N-6)$.

__Proof.__ If type (L,p) equals N, then some term in the expression in proposition 21 does not vanish. If this is the first term, we obtain immediately that $C(L,p) \leq N$. Otherwise, for the appropriate term, consider the inequality (as in the proof of the type 4 case) that

$$0 \leq |\lambda(L,\overline{\mathcal{L}}^{j+1})|^2 \leq \lambda(L,\overline{L})\lambda(\overline{\mathcal{L}}^{j+1},\mathcal{L}^{j+1}). \tag{23}$$

From this inequality, since $v_L\lambda(L,\overline{\mathcal{L}}^{j+1}) \leq N-2-j$, and using remark 3, we obtain the fact that

$$v_L(\lambda(L,\overline{L})) \leq 2(N-j-2)v_j. \tag{24}$$

(Note that when $j=0$, we get that $v_L(\lambda(L,\overline{L})) \leq N$, by adding v_0 to both sides of 24.) Otherwise, since $C(L,p) =$ (left side of 24) + 2, we obtain the desired formula. Since we are guaranteed that only one term does not vanish, we must take the maximum over j to conclude the desired formula. To improve it to $2N-6$, we first look at the case type $(L,p) = 6$.

__25. Example.__ Suppose that type (L,p) equals 6. The result of theorem 22 tells us that $C(L,p) \leq \max(8-v_1,6-v_2,4-v_3,2-v_4,6)$. However, we have already seen that, if type $(L,p) \geq 4$, then $v_1 \geq 2$. Therefore, we actually have that $C(L,p)$ is at most 6. Since it must be even, and, by theorem 13, it is not 4, we have that $C(L,p)$ equals 6.

This now enables us to obtain the final statement in the theorem, because the only term not already $\leq 2N-6$ is the term when $j=1$. By including the fact that v_1 is at least 2, the result follows. We believe that this process can be carried out further, to obtain

__26. Conjecture.__ If M is pseudoconvex, then type $(L,p) = C(L,p)$. More generally, the assumption of pseudoconvexity is only required on the span of all the $(1,0)$ parts of iterated commutators.

This conjecture would follow if we could prove the analogous statement for higher orders than four. What we would have to show is the following:

Suppose that $C(L,p) > N$, and that M is pseudoconvex. Then, for all possible choices of iterated commutators and resulting operators $\mathscr{L}^{j+1}$,

$v_L\lambda(\mathscr{L}^{j+1},\overline{\mathscr{L}}^{j+1})$ is at least $N-2j-2$. This must hold for $0 \leq j \leq N-1$. (27)

Thus the Levi forms of these (1,0) parts of iterated commutators must vanish to various orders themselves.

__28. Lemma.__ Suppose that M is pseudoconvex, and that 27 holds. Then we have that each term in the formula 21.1 vanishes at p, and hence we have that type $(L,p) \geq N$.

__Proof.__ This can be derived analogously to 24, or we can simply note that if the result failed to be true, and if v_j were $\geq N-2-2j$, the estimate in the conclusion of theorem 22 would give $C(L,p) \leq N$.

__29. Remark.__ If M is a real hypersurface of $\mathbb{C}^n$, and L is tangent to a complex manifold, then it is easy to see that the only terms in the contraction of an iterated bracket are those of the form $\mathscr{D}^j\lambda(L,\overline{L})$, so the result holds. For general vector fields, some pseudoconvexity hypothesis is required. Pseudoconvexity plays a subtle role here. See the examples in [1].

__Bibliography__

1. T. Bloom, "On the contact between complex manifolds and real hypersurfaces in $\mathbb{C}^3$", Trans. A. M. S., Vol. 263, No. 2, Feb. 1981, 515-529.

2. D. Catlin, "Boundary invariants of pseudoconvex domains", Annals of Math. 120 (1984), 529-586.

3. J. D'Angelo, "Finite type conditions for real hypersurfaces", J. Diff. Geom. 14 (1979), 59-66.

4. __________, "Points of finite type on real hypersurfaces", (preprint).

5. __________, "Real hypersurfaces, orders of contact, and applications", Annals of Math. 115 (1982), 615-637.

6. K. Diederich and J. Fornaess, "Pseudoconvex domains with real analytic boundary", Annals of Math. 107 (1978), 371-384.

7. J. J. Kohn, "Boundary behavior of $\overline{\partial}$ on weakly pseudoconvex manifolds of dim 2", J. Diff. Geom. 6 (1972), 523-542.

PLURISUBHARMONIC FUNCTIONS ON RING DOMAINS

by

John Erik Fornaess [1] and Nessim Sibony

1. Introduction.

Griffiths [G1] showed in 1971 that holomorphic maps from a punctured ball in $\mathbb{C}^n$, $n \geq 3$, to a compact Kähler manifold extend meromorphically across the puncture. Shiffman [S1] extended this result to $\mathbb{C}^2$ (See also Siu [SIU2] and Sibony [SIB1]). A key step in the proof was to pull back the Kähler form ω via the holomorphic map f. Then finiteness of the volume of the graph of f reduced to the estimate $\int (f^*\omega)^n < \infty$. Since there always is a C^∞ plurisubharmonic function u on the punctured ball $\mathbb{B}^*$ such that $dd^c u = f^*\omega$, the estimate reduced to whether $\int_{\mathbb{B}^*} (dd^c u)^n < 0$ for any C^∞ plurisubharmonic function u on $\mathbb{B}^*$ (shrinking the radius if necessary). This estimate was shown to hold.

Griffiths [G1] proposed to study extendability of holomorphic maps, into compact Kähler manifolds, across balls rather than isolated points. (By the Hartogs extension theorem this follows if the target is complex projective space). This leads to the question whether $\int (dd^c u)^n < \infty$ where K is a closed ball in $\mathbb{C}^n$, Ω is an open ball containing K and u is a smooth plurisubharmonic function on a neighborhood of $\overline{\Omega}$-K. We are far from able to do this. But we have made some observations about plurisubharmonic functions on ring domains $\Omega_K = \Omega$-K where K can also be a closed polydisc or a general polynomially convex compact set.

In section 2 we show integral estimates that are weaker than the above. We show for example that u is in $L^1(\Omega_K)$, see Theorem 2.2 for a precise statement.

Smoothing of plurisubharmonic functions on ring domains is discussed in section 3. We show that smoothing is always possible on a general class of ring domains. It is known that smoothing is impossible in some cases on smooth bounded domains in $\mathbb{C}^2$ ([F1]).

A plurisubharmonic function ρ on a ring domain $\Omega_K = \mathbb{B}$-K is said to have a plurisubharmonic subextension to $\mathbb{B}$ if there exists a plurisubharmonic function σ on $\mathbb{B}$, $\sigma \not\equiv -\infty$, such that $\sigma \leq \rho$ on Ω_K. By Bedford-Burns [B1] we cannot expect ρ to have a plurisubharmonic extension to $\mathbb{B}$. We don't know whether subextensions exist when ρ is smooth and K is a ball. But in section 4 we give an example of a smooth ρ which does not admit a subextension, K being a polydisc and in section 5 we describe a discontinuous ρ which does not admit a subextension, K being a ball.

1) Author has been supported in part by an NSF grant.

2. Integral estimates.

Let $\Omega_K = \mathbb{B} - K$ be a ring domain in $\mathbb{C}^n$, $n \geq 1$, and let ρ be a plurisubharmonic function on Ω_K. Our first observation is that ρ is bounded above in a neighborhood of K.

Lemma 2.1. *If K is polynomially convex, there exists a neighborhood V of K, $V \subset \mathbb{B}$, such that* $\sup_{z\in V-K} \rho(z) < \infty$.

Proof. Fix any open neighborhood V of K, $V \subset\subset \mathbb{B}$, and let $p \in V-K$. Then there exists a holomorphic polynomial $P(z)$ such that $P(p) = 1$ and $|P| < 1$ on K. Let Z be the complex analytic variety $Z = \{z \in \mathbb{C}^n;\ P(z) = 1\}$. Then $Z \cap K \neq \emptyset$ and so, by the maximum principle for plurisubharmonic functions,

$$\rho(p) \leqq \sup_{z\in Z\cap\partial V} \rho(z) \leqq \sup_{z\in\partial V} \rho(z) < \infty$$

It is clear from the proof that it suffices to assume that K is n-1 meromorphically convex, i.e. that for every point $p \in \mathbb{B} - K$ there exist holomorphic functions $f_1, \ldots, f_{n-1}$ on $\mathbb{B}$ such that $\sum_{i=1}^{n-1} |f_i| > 0$ on K and $f_i(p) = 0$, $i = 1, \ldots, n-1$.

Because of Lemma 2.1 we can without loss of generality restrict our considerations to negative plurisubharmonic functions.

Theorem 2.2 *Assume* $u < 0$ *is plurisubharmonic and smooth on a neighborhood of* $\overline{\mathbb{B}}$-K, *where K is a closed ball. Then*

$$\int_\Omega |u|(dd^c u)^{n-1} \wedge dd^c|z|^2 < \infty \text{ , and if}$$

$$K = \overline{\mathbb{B}}(0,r), \int_{\Omega_K} (|z|-r)\ (dd^c u)^n < \infty.$$

Observe that we can add a multiple of $|z|^2$ to u. Applying the first integral estimate to the new function it follows that u is in $L^1(\Omega_K)$.

Proof. Let $\rho = \max\{|z|^2 - r^2, 0\}$ and let $\{\rho_k\}$ be a sequence of non-negative smooth plurisubharmonic functions converging uniformly to ρ, and even in C^2 - norm on compact subsets of Ω_K, $\rho_k \equiv 0$ in some neighborhood of $\overline{\mathbb{B}}(0,r)$, $\rho_k \equiv \rho$ near $\partial\mathbb{B}$. Integrating by parts twice we get that

$$\int_{\Omega_K} \rho_k (dd^c u)^n = \int_{\partial\Omega_K} \rho_k d^c u \wedge (dd^c u)^{n-1}$$

$$-\int_{\Omega_K} d\rho_k \wedge d^c u \wedge (dd^c u)^{n-1} =$$

$$\int_{\partial\mathbb{B}} \rho d^c u \wedge (dd^c u)^{n-1} - \int_{\partial\mathbb{B}} u d^c \rho \wedge (dd^c u)^{n-1}$$

$$+ \int_{\Omega_K} u\, dd^c \rho_k \wedge (dd^c u)^{n-1}.$$

Hence

$$\int_{\Omega_K} \rho_k (dd^c u)^n + \int_{\Omega_K} |u| dd^c \rho_k \wedge (dd^c u)^{n-1}$$

is bounded by a constant independent of k. Since both integrands are nonnegative, the theorem follows.

Let u be plurisubharmonic on a ring domain Ω_K, K polynomially convex, $u \not\equiv -\infty$. Then $dd^c u$ is a closed, positive (1,1) current. Instead of asking for a plurisubharmonic subextension v of u we can ask for a closed, positive (1,1) current S on $\mathbb{B}$ which is a superextension of $dd^c u$, i.e. $S \geq dd^c u$ on Ω_k. We don't know if $dd^c u$ has a superextension whenever u has a subextension. However, the opposite is true. As indicated in the introduction we would like to know if the stronger estimate

$\int (dd^c u)^n < \infty$ holds.

<u>Proposition 2.3</u>. <u>If there exists a positive, closed</u> (1,1) <u>current</u> S <u>on</u> $\mathbb{B}$ <u>with</u> $S \geq dd^c u$ <u>on</u> Ω_K, u <u>plurisubharmonic on a neighborhood of</u> $\overline{\mathbb{B}}\setminus K$, <u>then there exists a plurisubharmonic subextension</u> v <u>of</u> u <u>to</u> $\mathbb{B}$, $v \leq u$ <u>on</u> Ω_K.

Proof. There exists a plurisubharmonic function ρ on $\mathbb{B}$ such that $S = dd^c\rho$([H1]). Then $dd^c(\rho - u) \geq 0$ and hence $\rho - u$ is plurisubharmonic on Ω_K. By Lemma 2.1 $\rho - u$ is bounded above on Ω_K. Hence v: $= \rho - c$ is a subextension of u if c is a large enough positive constant.

3. <u>Smoothing</u>.

We will prove a smoothing theorem for plurisubharmonic functions on ring domains Ω_K where K is a starshaped compact satisfying the following condition:

$$tK \subset \operatorname{int} K \text{ if } t \in \mathbb{R},\ 0 \leq t < 1. \tag{1}$$

<u>Theorem 3.1</u>. <u>Let</u> ρ <u>be plurisubharmonic on a ring domain</u> $\Omega_K = \mathbb{B} - K$ <u>where</u> K <u>is compact and satisfies condition</u> (1). <u>Then there exists a sequence</u> $\{\rho_n\}_{n=1}^{\infty}$ <u>of</u> C^∞ <u>plurisubharmonic functions such that</u> $\rho_n \searrow \rho$ <u>pointwise when</u> $n \to \infty$.

<u>Proof</u>. By the theorem of Richberg ([R1]) it suffices to find a sequence $\{\rho_n\}_{n=1}^{\infty}$ of <u>continuous</u> plurisubharmonic functions converging down to ρ.

Notice that if $\{\sigma_m\}_{m=1}^{\infty}$ is a sequence of plurisubharmonic functions converging down to σ and we can smooth each σ_m, then we can smooth σ. To see this, assume $\{\sigma_{m,n}\}_{n=1}^{\infty}$ are continuous and plurisubharmonic on Ω_K and $\sigma_{m,n} \searrow \sigma_m$ pointwise. Let $\{K_m\}_{m=1}^{\infty}$ be a sequence of compact sets in Ω_K, $K_m \subset \text{int } K_{m+1}$ $\forall$ m and $\Omega_k = \cup K_m$. Choose for each m an integer n(m) such that $\sigma_{m,n(m)} \le \sigma_{k,m} + \frac{1}{m}$ on K_m $k \le m$. Then $\tau_k = \sup_{m \ge k} \sigma_{m,n(m)}$ is continuous and plurisubharmonic on $\Omega_K | K$ and $\tau_k \searrow \sigma$.

Since $\rho^{(k)} = \max\{\rho, -k\}$ converge down to ρ, it suffices by the above observation to smooth ρ in the case ρ is bounded below.

Let $\{\mathbb{B}_m\}$ be a sequence of balls, $K \subset \mathbb{B}_1 \subset\subset \ldots \subset\subset \mathbb{B}_m \subset\subset \mathbb{B}_{m+1} \subset\subset \cup \mathbb{B}_m = \mathbb{B}$ and let σ_m be a continuous plurisubharmonic function, $\sigma_m \le \rho$ on $\mathbb{B}_m$-K and $\sigma_m \ge \rho$ on $\mathbb{B} - \mathbb{B}_{m+1}$.

Let $\rho^{(k)} := \sup\{\rho, \sup_{m \ge k} \sigma_m\}$. Then each $\rho^{(k)}$ is plurisubharmonic on Ω_k, $\rho^{(k)} \searrow \rho$ and $\rho^{(k)}$ is continuous on $\mathbb{B} - \bar{\mathbb{B}}_{k+1}$. Using again the same observation we can assume that ρ is bounded below and is also continuous near $\partial \mathbb{B}$. We may also assume that there exists a k and a plurisubharmonic function σ on $\mathbb{C}^n$-K such that ρ is continuous on $\mathbb{B} - \mathbb{B}_K$, $\sigma = \rho$ on $\mathbb{B}_{k+1}$-K and $\sigma < \rho$ on $\bar{\mathbb{B}}_{k+2} - \mathbb{B}_{k+1}$. It then suffices to smooth σ on $\mathbb{C}^n$-K, because if $\{\sigma_n\}$ is a smoothing sequence for σ, then

$$\rho_n := \begin{cases} \max\{\sigma, \rho\} \text{ on } \bar{\mathbb{B}}_{k+2}\text{-}K \\ \rho \text{ on } \mathbb{B} | \bar{\mathbb{B}}_{k+2} \end{cases}$$

is a smoothing sequence for ρ on Ω_k if we start with large enough n.

We assume then that ρ is plurisubharmonic on $\mathbb{C}^n$-K and that $\rho \not\equiv -\infty$. We will find a sequence of continuous plurisubharmonic functions converging down to ρ on $\mathbb{C}^n$-K.

Fix a C^∞ function $\chi: \mathbb{C}^n \to \mathbb{R}^+ \cup \{0\}$ with supp $\chi \subset \mathbb{B}(0,1)$. Let $t \in (0,1)$. By the star-shaped condition on K, $tK \subset$ int K. Hence there exists an $\varepsilon = \varepsilon(t) > 0$ such that $\varepsilon < 1-t$ and $\mathbb{B}(p,\varepsilon) \subset \mathbb{C}^n$-tK $\forall p \in \mathbb{C}^n$-K. We can assume that ε depends continuously on t and that $\varepsilon(t) \searrow 0$.

For $t \in (0,1)$ let ρ_t be the plurisubharmonic function on $\mathbb{C}^n$-tK defined by $\rho_t(z) = \rho_t(z) = \rho(z/t)$. Also, define a smoothing σ_t of ρ_t on $\mathbb{C}^n$-K by

$$\sigma_t(z) = \frac{1}{(\varepsilon(t))^{2n}} \int_{\omega \in \mathbb{B}(0,\varepsilon)} \chi\left(\frac{\omega}{\varepsilon}\right) \rho_t(z+\omega)$$

Then σ_t is C^∞, plurisubharmonic and $\sigma_t \geqq \rho_t$. Hence we obtain the crucial estimate

$$\sigma_t(z) \ge \rho(z/t). \tag{2}$$

If $k \ge 2$, let $\rho_k(z) = \sup\limits_{1-\frac{1}{k}<t<1} \sigma_t(z)$.

We show that each ρ_k is continuous and plurisubharmonic on $\mathbb{C}^n\setminus K$ and that $\rho_k \searrow \rho$ pointwise.

At first we show

$$\rho_k(z) \ge \rho(z)\ \forall\, k, \forall\, z\varepsilon\mathbb{C}^n-K. \tag{3}$$

This follows from (2) because

$$\rho(z) = \overline{\lim_{t\nearrow 1}}\ \rho(z/t), \text{ ([HE1]).}$$

Next we show that

$$\sigma(t,z) := -\ \sigma_t(z) \text{ is continuous on } (0,1) \times (\mathbb{C}^n-K). \tag{4}$$

To prove (4) it suffices to demonstrate continuity of

$$\sigma'(t,z) := \int\limits_{\omega\varepsilon\mathbb{B}(z,\varepsilon(t))} \chi\left(\frac{\omega-z}{\varepsilon(t)}\right)\ \rho\left(\frac{\omega}{t}\right)$$

but this is immediate since $\rho\varepsilon\ L^1_{\ell oc}$.

$$\overline{\lim_{(t,\omega)\to(1,z)}}\ \sigma(t,z) = \rho(z) \tag{5}$$

This follows since (3) gives $\ge$ and upper semicontinuity of ρ gives $\le$. From (3) and (5) and the obvious monotonicity it follows that $\rho_k \searrow \rho$. It only remains to show that each ρ_k is continuous. Plurisubharmonicity is then automatic since $\rho_k = \rho_k^*$, the upper regularization of ρ_k. Fix k, z and $\varepsilon > 0$. We show continuity of ρ_k at z. From (4) and (5) it follows that $\rho_k(z) < \infty$. From the definition of ρ_k and the smoothness of each σ_t it follows that $\rho_k(\omega) > \rho_k(z) - \varepsilon$ for all ω in some neighborhood of z. From (3) and (5) it follows that there exists a t_o, $1-1/_k < t_o < 1$, and a neighborhood U of z such that $\sigma_t(\omega) < \rho_k(z) + \varepsilon/2\ \forall\, \omega\varepsilon U$ and $t_o \leqq t < 1$. Shrinking U if necessary we obtain from (4) the same inequality for $\omega\varepsilon U$, $1-1/k < t \leqq t_o$. Hence $\rho_k(\omega) < \rho_k(z) + \varepsilon \ \forall \ \omega\varepsilon U$ and continuity follows.

We don't know if we can smooth on domains $\Omega_1\setminus\bar{\Omega}_2$ if $\Omega_2 \subset\subset \Omega_1$ are two smoothly bounded strongly pseudoconvex domains.

4. Subextension across a polydisc.

We will prove the following theorem.

Theorem 4.1. There exists a C^∞ plurisubharmonic function u on $\mathbb{B}^2(0,2) - \bar{\Delta}^2(0,1)$ which does not have a plurisubharmonic subextension to $\mathbb{B}^2(0,2)$.

Proof. Choose first a function $g\varepsilon A^\infty(\{|z|\geq 1\})$ with the property that the formal power series of g diverges at every boundary point of the unit disc.

We may assume that $|g| < 1/3$. Next, choose a nonempty, perfect compact set E in $\{|z|=1\}$ which is a zero-set for an A^∞ function f on $\{|z|\geqq 1\}$, ([Cl]).

Let $\rho(z,w) = \log(|f(z)|^2 + |w-g(z)|^2)$ if $|z| > 1$. Choose a C^∞ function $\chi(w)$ with $\chi(w)\equiv 1$ if $|w| < 1/2$ and $\chi(w)\equiv 0$ if $|w| > 2/3$. If K is large enough, then u: = $\chi(w)\ \rho(z,w) + K|(z,w)|^2$ is plurisubharmonic on $\mathbb{B}^2(0,2) - \bar{\Delta}^2(0,1)$.

Assume that u has a plurisubharmonic subextension σ across $\bar{\Delta}^2(0,1)$. Then the Lelong-number $\nu(z,w)$ of σ at every point $(z,g(z))$, $z\varepsilon E$, is at least one. By Siu's theorem ([SIU 1]) the set $X = \{(z,w);\ \nu(z,w) \geq 1\}$ is a complex-analytic set. Except for finitely many points in $E = \{(z,g(z));\ z\epsilon E\}$, every point in $\tilde{E}$ is a regular point of X and X is locally a graph over the z-axis of a holomorphic function. This implies that g has a convergent formal power series at every point in E except finitely many. Hence we have a contradiction to our choice of g.

Remark 4.2. It is impossible to approximate u on compact subsets by plurisubharmonic functions on $\mathbb{B}(0,2)-\bar{\Delta}^2(0,1-\delta)$, $\delta > 0$ fixed.

If $u_n \to u$ on compact sets, then $v: = (\overline{\lim_{n\to\infty}}\, u_n)^*$ is a plurisubharmonic subextension of u to $\mathbb{B}(0,2)-\bar{\Delta}^2(0, 1-\delta)$.

For this remark it would suffice to use the nonextendable functions constructed by Bedford-Burns ([Bl]).

5. Subextension across a ball.

Theorem 5.1. There exists a plurisubharmonic function ρ on $\mathbb{B}^3(0,2)-\bar{\mathbb{B}}^3(0,1)$, $\rho \not\equiv -\infty$ which does not admit a plurisubharmonic subextension to $\mathbb{B}^3(0,2)$.

We don't know whether ρ can be chosen to be smooth. We also don't know whether ρ can be chosen such that the trivial (=0) extension across $\bar{\mathbb{B}}^3(0,1)$ has infinite Lelong number at one boundary point of $\bar{\mathbb{B}}^3(0,1)$.

The proof will show that the restriction of ρ to $\mathbb{B}^3(0,b) - \mathbb{B}^3(0,a)$ fails to have a subextension to $\mathbb{B}^3(0,b)$ for any a, b, $1< a < b < 2$.

Let $\alpha_1,\ldots,\alpha_n$ be distinct complex numbers, and let $0 < \varepsilon << \delta << 1$. We will at first show:

Lemma 5.2. *The set* $X_{\varepsilon,\delta}$ *in* $\mathbb{C}^3(z,\omega,\eta)$, $X_{\varepsilon,\delta} = \{\prod_{i=1}^{n} (\omega-\alpha_i z) = \varepsilon \wedge \prod_{i=1}^{n}(\eta-\alpha_i z)=\delta\}$ *is an irreducible complex manifold.*

Proof of Lemma 5.2. Let $X_\varepsilon = \{\prod_{i=1}^{n} (\omega-\alpha_i z) = \varepsilon\}$, $\varepsilon > 0$. Then X_ε is a complex manifold in $\mathbb{C}^2$ since $\nabla(\prod_{i=1}^{n} (\omega-\alpha_i z)) \neq 0$ everywhere except at the origin.

We show that X_ε is connected. First, observe that X_ε is an n-sheeted branched covering over $\mathbb{C}(z)$ with branch points $a_1^\varepsilon,\ldots,a_k^\varepsilon \in \mathbb{C}(z)$. All branch points are non-zero. Let $\eta^\varepsilon = \sup_j \{|a_j^\varepsilon|\}$. Then $\eta^\varepsilon \to 0$ as $\varepsilon \to 0$. When $|z| > \eta^\varepsilon$, X_ε consists of n disjoint graphs, $\{X_\varepsilon^j\}$, $j = 1,\ldots,n$, $X_\varepsilon^j = \{\omega = f_\varepsilon^j(z)\}$, $f_\varepsilon^j(z) - \alpha^j z \to 0$ when $z \to \infty$ if $n \geq 2$. If X_ε has two components, X_ε^1 and X_ε^2, let $(z, \delta^1(z)),\ldots,(z, \delta^p(z))$ and $(z, \gamma^1(z)),\ldots, (z, \gamma^q(z))$ locally parametrize, with multiplicities, the two components. Then $f(z) := \prod_{\substack{i=1,\ldots,p \\ j=1,\ldots,q}} (\delta^i(z)-\gamma^j(z))$ is a polynomial of degree pq without zeroes. This is of course impossible, so X_ε is connected.

Next we fix $0 < \varepsilon << \delta << 1$ so that $\{a_j^\varepsilon\} \cap \{a_j^\delta\} = \emptyset$. We may assume that $\eta^\varepsilon, \eta^\delta < 1$. Let $R_j(S_j)$ be a closed line segment starting at a_j^ε (a_j^δ) and ending on the unit circle, $j = 1,\ldots,k$. We can choose the line segments to be pairwise disjoint and so that 1 is on none of them. Let $\{(1, \zeta_i,\xi_j)\}_{i,j=1}^n$ be the points in $X_{\varepsilon,\delta}$ with $z = 1$.

We consider $X_{\varepsilon,\delta}$ as a branched cover over the z-axis. The branch points in $\mathbb{C}(z)$ are $\{a_j^\varepsilon\}_{j=1}^k \cup \{a_j^\delta\}_{j=1}^k$.

We show that $X_{\varepsilon,\delta}$ is a complex manifold. If $(z_o, w_o, \eta_o) \in X_{\varepsilon,\delta}$ we consider several cases. If (z_o, w_o) is not a branch point for X_ε and (z_o, η_o) is not a branch point for X_δ then we can parametrize $X_{\varepsilon,\delta}$ as $\{(z, f(z), g(z))\}$ nearby, f, g holomorphic. If (z_o, w_o) is a branch point for X_ε, then $z_o = a_j^\varepsilon$ for some j and X_δ can be parametrized as $\{(z, g(z))\}$ nearby. In that case X_ε can be parametrized as $\{(f(\omega), \omega)\}$, hence $X_{\varepsilon,\delta}$ can be described by $\{(f(\omega), \omega, g(f(\omega)))\}$. A similar argument applies when (z_o, η_o) is a branch point for X_δ.

To show that $X_{\varepsilon,\delta}$ is irreducible, it suffices to find curves in $X_{\varepsilon,\delta}$ connecting any two points in $\{(1, \zeta_i, \xi_j)\}$. Fix a ξ_j and ζ_{i_1}, ζ_{i_2}. By summetry it will suffice to find a curve connecting $(1, \zeta_{i_1}, \xi_j)$ to $(1, \zeta_{i_2}, \xi_j)$.

Let γ be a curve in X_ε from ζ_{i_1} to ζ_{i_2}. We can assume that the projection of γ to the z-plane, λ, avoids all branch points $\{a_j^\varepsilon\}$, $\{a_j^\delta\}$. Then λ has a unique lifting Λ to $X_{\varepsilon,\delta}$ which starts at $(1, \zeta_{i_1}, \xi_j)$. Also, Λ ends at $(1, \zeta_{i_2}, \xi_\ell)$ for some ℓ. If $\ell \neq j$, we will change γ so that Λ ends at $(1, \zeta_{i_2}, \xi_j)$. We may at first assume that λ is contained in the closed unit disc and consists of finitely many arcs on the unit circle and finitely many straight lines. We can assume that λ only intersects R_j's or S_j's on the unit circle. For every corner of λ in the open unit disc add and subtract a smooth curve going to the unit circle without intersecting any R_j or S_j. Subdividing the straight line segments further at first we can assume that λ has the following property: λ consists of finitely many arcs on the unit circle and finitely many curves $\{\lambda_i\}$ starting and ending on the unit circle. Furthermore each λ_i can be contracted to the unit circle T or to $T \cup S_j$ or to $T \cup R_j$ for some j without intersecting any (other) S_k or R_k. If λ_i contracts to R, replace λ_i by such a contraction. If λ_i contracts to $T \cup S_j$, contract λ_i further to T. These last contractions change Λ so that Λ ends up at $(1, \zeta_{i_2}, \xi_j)$. Hence we have shown that $X_{\varepsilon,\delta}$ is irreducible.

Observe that if $\varepsilon = \delta$, $n \geq 2$, then $X_{\varepsilon,\delta}$ is reducible. One component is given by the equations $\{ \prod_{i=1}^{n}(\omega - \alpha_i z) = \varepsilon$ and $\eta = \omega\}$.

<u>Proof of Theorem 5.1</u>. For each integer $m >> 1$ choose distinct complex numbers $\alpha_1^m, \dots, \alpha_{N(m)}^m$ such that for all $\varepsilon_m, \delta_m > 0$ small enough, all points in $\mathbb{B}^3(0,2) - \bar{\mathbb{B}}^3(0,1)$ are closer to $X_m := \{\prod_{i=1}^{N(m)}(\omega - \alpha_i^m z) = \varepsilon_m \wedge \prod_{i=1}^{N(m)}(\eta - \alpha_i^m z) = \delta_m\}$ than $\exp(-m^3)$. We may assume that $\alpha_1^m = 0$ for all m and that $\varepsilon_m << \delta_m << 1$ as in Lemma 5.2. We also assume that ε_m, δ_m are small enough for the following to hold. We can describe X_m near $\{\eta = \omega = 0\}$ as a graph, $\eta = g_m(z)$, $\omega = f_m(z)$, valid for $\frac{1}{2} < |z| < 3$. Set $\phi_m(z, \omega, \eta) =$

$$\frac{1}{m^2}\ell n(|\omega - f_m(z)|^2 + |\eta - g_m|^2) \text{ and } \psi_m(z, \omega, \eta) = \frac{1}{m^2}\ell n(|\omega|^2 + |\eta|^2) + \frac{1}{m^2}(|\omega|^2 + |\eta|^2) - \frac{1}{m^4}.$$

Define a plurisubharmonic h_m on $\mathbb{B}(0,2) - \bar{\mathbb{B}}(0,1)$ by

$$h_m = \begin{cases} \phi_m \text{ if } |\omega|^2 + |\eta|^2 < \frac{1}{2m^2} \\ \max\{\phi_m, \psi_m\} \text{ if } \frac{1}{2m^2} \leqq |\omega|^2 + |\eta|^2 \leqq \frac{2}{m^2} \\ \psi_m \text{ if } |\omega|^2 + |\eta|^2 > \frac{2}{m^2} \end{cases}$$

Let $\rho := \Sigma h_m$. Then ρ is plurisubharmonic on $\mathbb{B}(0,2) - \overline{\mathbb{B}}(0,1)$, $\rho \not\equiv -\infty$. We will show that ρ has no subextension.

Assume that σ is plurisubharmonic on $\mathbb{B}^3(0,2)$, $\sigma \not\equiv -\infty$ and $\sigma \leqq \rho$ on $\mathbb{B}^3(0,2)-\overline{\mathbb{B}}^3(0,1)$. Fix an m. Let Y_m consist of the points in $\mathbb{B}(0,2)$ where the Lelong-number of $\rho \geq 1/m^2$. By Siu's theorem ([SIU1]), Y_m is a complex analytic variety and by construction, Y_m contains a nonempty relatively open subset of X_m. Hence Y_m contains X_m.

If μ is plurisubharmonic on $\mathbb{B}(0,r)$, $\mu \leqq K$ and the Lelong-number of μ at 0 is at least $c > 0$, then $\mu \leqq c \log|z| + K = c\log r$. Fix a point p in $\mathbb{B}(0,2) - \overline{\mathbb{B}}(0,1)$. Then there exist K, $r > 0$ such that for all large enough m,

$$\sigma(p) \leqq \frac{1}{m^2} \log(\exp(-m^3)) + K - \frac{1}{m^2} \log r$$

Hence $\sigma \equiv -\infty$ contrary to assumptions.

Remark 5.3. As follows from proposition 2.3 the current $T = dd^c\rho$ does not have a superextension as a closed current in $\mathbb{B}^3(0,2)$. But if S is the current of integration on an analytic hypersurface in $\mathbb{B}^3(0,2)-\overline{\mathbb{B}}^3(0,1)$, then S extends to a closed current in $\mathbb{B}^3(0,2)$ (because the variety extends).

It is easy to construct a similar example in $\mathbb{B}^2(0,2)-\overline{\mathbb{B}}^2(0,1)$. In that case the analytic sets (of dimension one) do not extend in general.

References

[B1] Bedford, E., Burns, D.: Domains of existence for plurisubharmonic functions, Math. Ann. 238 (1978), 67-69.

[C1] Carleson, L.: Sets of uniqueness for functions regular in the unit circle.' Acta Math. 87 (1952), 325-345.

[F1] Fornæss, J.E.: Plurisubharmonic functions on smooth domains, Math. Scand. 53 (1983), 33-38.

[G1] Griffiths, P.: Two theorems on extension of holomorphic mappings, Inv. Math 14 (1971), 27-62.

[H1] Harvey, R., Shiffman B.: A characterization of holomorphic chains, Ann. Math. 99 (1974), 553-587.

[HE1] Helms, L.: Introduction to potential theory, Wiley-Interscience (1969).

[R1] Richberg, R.: Stetige streng pseudokonvexe Funktionen, Math. Ann. 175 (1968), 251-268.

[S1] Shiffman, B.: Extension of line bundles and meromorphic maps, Inv. Math. 15 (1972), 332-347.

[SIB1] Sibony, N.: Quelques problèmes de prolongement de courants en analyse complexe, Duke Math. J. 52 (1985), 157-197.

[SIU1] Siu, Y.T.: Analyticity of sets associated to Lelong numbers and the extension of closed positive currents, Inv. Math. 27 (1974), 53-156.

[SIU2] Siu, Y.T.: Extension of meromorphic maps to Kähler manifolds, Ann. Math. 102 (1975), 421-462.

Characterizations of Certain Weakly Pseudoconvex Domains with Non-Compact Automorphism Groups

by

Robert E. Greene*

and

Steven G. Krantz*

*Work supported in part by the National Science Foundation

INTRODUCTION

Let $\Omega \subset\subset \mathbb{C}^n$ be a smoothly bounded domain. Let Aut Ω be the group of biholomorphic self-maps (or automorphisms) of Ω. It is a striking result of Bun Wong [30] that if Ω is strongly pseudoconvex and Aut Ω is non-compact then Ω is biholomorphic to the ball. More recently Rosay [26] has shown that if Ω is a domain, if there is a $Q_0 \in \partial\Omega$ near which $\partial\Omega$ is C^2 and strongly pseudoconvex, and if there is a $P_0 \in \Omega$ and $\varphi_j \in \text{Aut}\ \Omega$ such that $\varphi_j(P_0) \to Q_0$, then Ω is biholomorphic to the ball. (Observe that in $\mathbb{C}^1$ the result follows from standard uniformization techniques; new methods are needed when $n > 1$.)

It is the purpose of the present paper to explore what happens if, in the last sentence, Q_0 is a weakly pseudoconvex boundary point. The question cannot be answered in general at this time. But in case Q_0 is a special sort of "weak type" point in the sense of Kohn [18] and Bloom-Graham [3], then complete results can be obtained (see Theorems 1.1 and 1.2 below). A full understanding of the general type of boundary point Q_0 will involve much more subtle techniques. A discussion of what lies ahead is in Section 14.

Section 1 formulates the principal result of the paper: Theorems 1.1 and 1.2. Section 2 introduces the concepts and notation which are needed in the proofs. Section 3 records certain facts about analytic ellipsoids.

Section 4 collects the three main technical lemmas (Lemmas 4.1, 4.2, 4.3) which are needed in the proofs of Theorems 1.1. and 1.2. Lemma 4.1 is proved in Section 6, Lemma 4.2 is proved in Section 7, and the more difficult proof of Lemma 4.3 is contained in Sections 8-13.

Section 5 gives the proofs of Theorems 1.1 and 1.2, assuming the truth of Lemmas 4.1, 4.2 and 4.3.

The technical lemmas 4.1-4.3 should have independent interest for further investigations into mapping problems. Lemma 4.1 gives a method for constructing a biholomorphic mapping as a limit of a sequence of "almost" biholomorphic mappings. Lemma 4.2 gives a technique for showing that two domains are biholomorphically inequivalent. Lemma 4.3 reveals how certain biholomorphic invariants depend on boundary data. Also Sections 8 through 12 each contain useful results about dependence of biholomorphic invariants on boundary data.

The contents of Sections 5-14 are as follows:

Section 5. The Proofs of the Main theorems
Section 6. The Proof of Lemma 4.1
Section 7. The Proof of Lemma 4.2
Section 8. Uniform Estimates for the $\bar{\partial}$ Equation
Section 9. Analysis of Peaking Functions
Section 10. Uniform Approximation of Holomorphic Functions
Section 11. Localization of the Caratheodary and Eisenman Volume Forms
Section 12. Estimates on the Biholomorphic Invariant Q_Ω
Section 13. The Proof of Lemma 4.3
Section 14. Further Remarks and Speculations

It is a pleasure to thank S. Bell, J.E. Fornaess, S. Pincuk, and H.H. Wu for helpful discussions regarding the general subject matter of this paper.

The second author thanks Princeton University and the Institute for Advanced Study for their hospitality during a portion of this work.

§1. Statements of Results

This section presents the statements of the principal results. An explanation of their particular form is contained in Sections 3.

<u>Theorem 1.1</u>: Let $0 < m \in \mathbb{Z}$. Let $E_m \subseteq \mathbb{C}^2$ be given by

$$E_m = \{(z_1,z_2) : |z_1|^2 + |z_2|^{2m} < 1\}.$$

Let $\Omega \subset\subset \mathbb{C}^2$ be a domain with C^3 boundary such that

(i) $\mathbb{1} = (1,0) \in \partial\Omega$

(ii) There are neighborhoods U of $\mathbb{1}$ in $\partial\Omega$ and V of $\mathbb{1}$ in $\partial\Omega_m$ such that, up to a local biholomorphism, $U \cap \partial\Omega$ and $V \cap \partial E_m$ ooinoide,

If there are a point $P_0 \in \Omega$ and automorphisms φ_j of Ω such that $\varphi_j(P_0) \to \mathbb{1}$ as $j \to \infty$ then Ω is biholomorphic to E_m.

<u>Remark</u>: Notice that when $m = 1$ in Theorem 1.1 then the result becomes a special case of Rosay's theorem (of course Rosay does not require that $\partial\Omega$ be C^3, but that extra assumption is necessary for our methods).

The theorem is true in much greater generality than is stated. In fact, as the proof shows, hypothesis (ii) can be weakened considerably. What we actually need to assume about $\partial\Omega$ near $\mathbb{1}$ is that it is qualitatively like ∂E_m near $\mathbb{1}$. For example, it suffices to assume that, after a local biholomorphic change of coordinates near $\mathbb{1}$, $\partial\Omega$ satisfies

(ii') $\partial\Omega$ and ∂E_m osculate to order $2m + 1$ at points of the form $(e^{i\theta},0)$;

and

(ii") The eigenvalue of the Levi form has size $\sim|z_2|^{2m-1}$ at points of the form $(e^{i\theta},z_2)$.

In practice (ii') can often be arranged with a conformed mapping in the z_1 variable, once (ii") has been checked.

The analogue for Theorem 1.1 when $n > 2$ can be proved by the techniques of the present apper only in a special case. The result is now formulated as a separate theorem:

Theorem 1.2: The result of Theorem 1.1 holds for $\Omega \subseteq \mathbb{C}^n$, $n > 2$, provided Ω has C^{n+1} boundary and E_m has the form

$$E_m = \{z \in \mathbb{C}^n : |z_1|^2 + \ldots + |z_{n-1}|^2 + |z_n|^{2m} < 1\}.$$

§2. Notation and Definitions

Throughout this paper, the symbol Ω will denote a domain in $\mathbb{C}^n$ - that is, a bounded, connected open set. The domain Ω will be said to have C^k boundary if there is a defining function ρ for Ω which has the following special properties:

(i) $\rho : \mathbb{C}^n \to \mathbb{R}$,
(ii) ρ is C^k,
(iii) $\Omega = \{z \in \mathbb{C}^n : \rho(z) < 0\}$,
(iv) $\nabla\rho \neq 0$ on $\partial\Omega$.

Sometimes it is useful to define ρ only in a neighborhood of some boundary point and this will be done without comment.

Let $B = B_n \subseteq \mathbb{C}^n$ denote the unit ball, $\Omega(B)$ the holomorphic mappings from B into Ω, and $B(\Omega)$ the holomorphic mappings from Ω to B. If $F : \Omega_1 \to \Omega_2$ is a holomorphic mapping and $z \in \Omega_1$, let $F'(z)$ denote the Jacobian matrix (over $\mathbb{C}$) of F at z. In what follows, the symbol $|\xi|$ (without subscripts or superscripts) denotes Euclidean absolute value of the complex number ξ or Euclidean length of the vector $\xi \in \mathbb{C}^n$, depending on the context.

Now if $\Omega \subseteq \mathbb{C}^n$ is a domain and $z \in \Omega$ then the Caratheodary volume element of Ω at z is defined to be

$$M_\Omega^C(z) = \sup\{|\det F'(z)| : F \in B(\Omega), F(z) = 0\}.$$

Likewise the Eisenman volume element of Ω at z (sometimes called the Eisenman-Kobayashi volume element) is defined by

$$M_\Omega^E(z) = \inf\{1/|\det F'(0)| : F \in \Omega(B), F(0) = z\}.$$

It is immediate from the definitions that $M_\Omega^E(z) \geq M_\Omega^C(z)$ for all $z \in \Omega$. Also if $\varphi : \Omega_1 \to \Omega_2$ is holomorphic then $M_{\Omega_1}^E(z) \geq M_{\Omega_2}^E(\varphi(z))$ and $M_{\Omega_1}^C(z) \geq M_{\Omega_2}^C(\varphi(z)))$ for all $z \in \Omega_1$. It follows that the quotient

$$Q_\Omega(z) \equiv \frac{M_\Omega^E(z)}{M_\Omega^C(z)}$$

is a biholomorphic invariant in the sense that if $\varphi : \Omega_1 \to \Omega_2$ is biholomorphic then $Q_{\Omega_1}(z) = Q_{\Omega_2}(\varphi(z))$ for all $z \in \Omega_1$. See [22] for further details on these matters.

For completeness, and since they will be used occasionally later, the definitions of Caratheodary and Kobayashi metrics are reviewed here.

Let Ω be a domain, $z \in \Omega$, $\xi \in \mathbb{C}^n$. The infinitesimal form of the Caratheodary metric for Ω at z in the direction ξ is

$$F_\Omega^C(z,\xi) \equiv \sup\{|F'(z)(\xi)| : F \in B(\Omega), F(z) = 0\}.$$

Here $| \, |$ represents Euclidean length.

The Caratheodary distance between points $z, w \in \Omega$ is defined to be

$$\text{dist}_{\text{Car}}^\Omega(z,w) \equiv \sup_{f \in B(\Omega)} \rho(f(z), f(w)),$$

where ρ is the Poincaré-Bergman metric on B. This distance is not, in general, the same as that gotten by integrating F_Ω^C along curves. If $e_1 \equiv (1,0,\ldots,0)$ then the infinitesimal form of the Kobayashi metric for Ω at z in the direction ξ is

$F_\Omega^K(z,\xi) = \inf\{\alpha : \alpha > 0$ and $\exists F \in \Omega(B)$ such that
$F(0) = z$ and $F'(0)(e_1) = \xi/\alpha\}$.

Arc length and distance are constructed from F_Ω^K in the usual way (see [22]). The Kobayashi distance in Ω from z to w is denoted $\mathrm{dist}_{Kob}^{\Omega}(z,w)$. Holomorphic mappings are distance decreasing in both the Caratheodary and Kobayashi metrics. Biholomorphic mappings are isometries.

The following lemma, first discoverd by Bun Wong but proved in its present generality by Rosay, is the philosphical starting point for what is done in this paper. Again see [22] for further details.

Lemma 2.1: Let $\Omega \subset\subset \mathbb{C}^n$ be a domain (not necessarily with smooth boundary). There is a $z \in \Omega$ such that $Q_\Omega(z) = 1$ if and only if Ω is biholomorphic to the ball.

The other initial fact about M_Ω^C and M_Ω^E which will be needed here follows from the work of Graham [8]. Graham studies the asymptotic behavior of the infinitesimal form of the Caratheodary and Kobayashi metrics at points $z \in \Omega$ which approach strongly pseudoconvex boundary points. It was noted in [13] that the same calculations yield estimates for the asymptotic behavior of M_Ω^C, M_Ω^E near a strongly pseudoconvex boundary point. These ideas will now be reviewed, beginning with the notion of strictly pseudoconvex.

Definition 2.2: If $\Omega \subset\subset \mathbb{C}^n$ is a domain and $P \in \partial\Omega$ then P is called a point of (Levi) pseudoconvexity if there is a C^2 defining function for Ω near P satisfying

$$\sum_{j,k=1}^{n} \frac{\partial^2 \rho}{\partial z_j \partial \bar{z}_k}(P) w_j \bar{w}_k \geq 0 \tag{2.2.1}$$

for all complex tangent vectors $w = (w_1, \ldots, w_n)$ at P. Here a complex tangent vector is one satisfying

$$\sum_{j=1}^{n} \frac{\partial \rho}{\partial z_j}(P) w_j = 0.$$

The point P is said to be strongly (or strictly) pseudoconvex if

$$\sum_{j,k=1}^{n} \frac{\partial^2 \rho}{\partial z_j \partial \bar{z}_k}(P) w_j \bar{w}_k > 0 \tag{2.2.2}$$

for all non-zero complex tangent vectors w.

The definitions in 2.2 are easily seen to be independent of the choice of defining function. Indeed the quadratic form (2.2.1), called the Levi form, is the same for all local defining functions ρ which satisfy $|\nabla\rho(P)| = 1$. When reference is made below to the "eigenvalues of the Levi form at P", these will always be computed with respect to a defining function ρ which satisfies $|\nabla\rho(P)| = 1$. If $P \in \partial\Omega$ is a strongly pseudoconvex point then there exists a defining function $\tilde{\rho}$ for Ω near P such that

$$\sum_{j,k=1}^{n} \frac{\partial^2\tilde{\rho}}{\partial z_j \partial \bar{z}_k}(P)w_j\bar{w}_k \geq C|w|^2 \tag{2.3}$$

for all $w \in \mathbb{C}^n$. If $E \subseteq \partial\Omega$ is a compact subset of strongly pseudoconvex points then there are a $\tilde{\rho}$ and a constant C so that (2.3) is true simultaneously for all $P \in E$ and all $w \in \mathbb{C}^n$. See [22] for details.

If all points of $\partial\Omega$ are (strongly) Levi pseudoconvex then Ω is termed (strongly) pseudoconvex. Recall that C^2 Hartogs pseudoconvex domains are Levi pseudoconvex and conversely.

In all that follows, when analysis is being done near a strongly pseudoconvex boundary point, it will always be supposed that a defining function has been selected which satisfies (2.3).

Now the consequence of the calculation of Graham [8] which is of greatest interest here may be formulated as follows. (See [13], Corollary to Theorem 3.).

Proposition 2.4: Let $\Omega \subset\subset \mathbb{C}^n$ be strongly pseudoconvex. Let $\varepsilon > 0$. There is a $\delta > 0$ such that if $z \in \Omega$ and $\mathrm{dist}(z,\partial\Omega) < \delta$ then $1 \leq Q_\Omega(z) < 1+\varepsilon$.

One of the principal thrusts of the calculations in the present paper is to see how δ in Proposition 2.4 may be estimated in terms of ε and certain boundary data of Ω.

A final notion that will be needed is a non-isotropic "distance" modeled on the geometry of the Levi form near a strongly pseudoconvex boundary point. Let $\Omega \subset\subset \mathbb{C}^n$ have C^2 boundary. Let $P \in \partial\Omega$ and denote by ν_P the unit outward normal to $\partial\Omega$ at P. Define

$$\mathcal{N}_P = \mathbb{C}\nu_P.$$

and

$$\mathcal{T}_P = (\mathcal{N}_P)$$

to be the (Hermitian) orthogonal complement to $\mathcal{N}_P$.

If $z \in \bar{\Omega}$ then write

$$z = P + z_N + z_T, \quad z_N \in \mathfrak{N}_P, \quad z_T \in \mathcal{T}_P.$$

Such a decomposition, over $\mathbb{C}$, is unique. The vectors z_T and z_N are understood to be <u>defined</u> by the equation. Set

$$d_P(z) \equiv |z_N| + |z_T|^2.$$

This function d_P can be considered to measure distance, but it is in no sense a metric because i) it is defined only for $P \in \partial\Omega$ and ii) even when z and P both lie in $\partial\Omega$ it is not symmetric. The expression $d_P(z)$ will be a useful notational device in the technical calculations of Sections 6-11.

§3. Analytic Ellipsoids

If $m = (m_1, \ldots, m_n)$ is an n-tuple of positive integers, define the domain $\Omega = E_m$ to be

$$E_m = \{z \in \mathbb{C}^n : |z_1|^{2m_1} + \ldots + |z_n|^{2m_n} < 1\}.$$

There is no loss of generality to always suppose that $m_1 \le m_2 \le \ldots \le m_n$ and this will be done below without comment.

Of course when $m = (1, \ldots, 1)$ then E_m is the unit ball. For $m \ne (1, \ldots, 1)$ then E_m will not be strongly pseudoconvex precisely at points $(z_1, \ldots, z_n)$ for which there is a j satisfying $z_j = 0$ and $m_j > 1$. This follows from direct calculation, the details of which are omitted. If $m_{k+1}, \ldots, m_n$ are the indices which exceed unity then the weakly pseudoconvex points in ∂E_m consist of the union of $n-k$ real ellipsoids of dimension $2n - 3$; each of these ellipsoids intersects each of the others, and with transversal crossings.

If $\Omega \subset\subset \mathbb{C}^n$ is a domain, define

<u>Property K</u>: If $z_0 \in \Omega$, $\varphi_j \in \mathrm{Aut}\,\Omega$, and $\varphi_j(z_0) \to P \in \partial\Omega$ then for any $K \subset\subset \Omega$ it holds that $\varphi(z) \to P$ uniformly for $z \in K$.

That $\Omega = E_m$ satisfies Property K follows by a standard argument using the fact that each point of E_m has a holomorphic peaking function (see [22], Section 10.2).

A closely related phenomenon to Property K is Condition W for a domain Ω:

<u>Condition W</u>: There is a point $z_0 \in \Omega$, a sequence $\varphi_j \in \operatorname{Aut}\Omega$, and a $P \in \partial\Omega$ such that $\varphi_j(z_0) \to P$.

It is possible in principle for Condition W to hold without Property K (e.g. when P is a point of strong pseudoconcavity). Conversely, there may be no z_0 satisfying Property W, hence Property K would then hold vacuously.

If Ω is strongly pseudoconvex and Condition W holds then (by Bun Wong's theorem) Ω is the ball. However Condition W can hold when $P \in \partial\Omega$ is a weakly pseudoconvex point. Indeed, as will be seen in a moment, it holds for $\Omega = E_m$ if and only if $m_1 = 1$. If all $m_j > 1$ then $\operatorname{Aut} E_m$ contains only rotations and permutations of variables z_j, z_k with $m_j = m_k$. The proofs of these assertions about E_m are now given.

First, if $m_1 = 1$ then for $a \in \mathbb{C}$, $|a| < 1$, the map

$$\psi_a : (z_1,\ldots,z_n) \mapsto \left(\frac{z_1 - a}{1-\bar{a}z_1}, \frac{(1-|a|^2)^{1/2m_2} z_2}{(1-\bar{a}z_1)^{1/m_2}}, \ldots, \frac{(1-|a|^2)^{1/2m_n} z_n}{(1-\bar{a}z_1)^{1/m_n}} \right)$$

is an automorphism of E_m. Here $\operatorname{Re}(1-\bar{a}z_1) > 0$ for $z \in E_m$ so that the principal branch of logarithm may be defined. Now the collection of automorphisms $\varphi_j = \psi_{-1+1/j}$, the point $P = (1,0,\ldots,0)$, and (say) $z_0=0$ satisfy condition W.

For the converse, notice that if $m_1 > 1$ and condition W is satisfied on E_m with some family $\{\varphi_j\} \subseteq \operatorname{Aut} E_m$ and some $P = (P_1,\ldots,P_n) \in \partial E_m$ then it must be that some $P_j = 0$. For otherwise P would be a point of strong pseudoconvexity and Bun Wong's theorem would imply that E_m is biholomorphic to the ball. Say for simplicity that $P_n = 0$, $P_1 \neq 0$. Let

$$S = \{z \in \partial E_m : z_1 = 0\}.$$

Then S is a real $2n-3$ dimensional ellipsoid of weakly pseudoconvex points in ∂E_m which is disjoint from P. Now all automorphisms of E_m are covering transformation of the ball in the sense that if $\varphi \in \operatorname{Aut} E_m$ then there is a $\tilde{\varphi} \in \operatorname{Aut} B$ such that the diagram

$$\begin{array}{ccc} E_m & \xrightarrow{\varphi} & E_m \\ {\scriptstyle T_m}\downarrow & & \downarrow{\scriptstyle T_m} \\ B & \xrightarrow{\tilde{\varphi}} & B \end{array}$$

commutes, where

$$T_m : E_m \to B$$

$$(z_1,\dots,z_n) \mapsto (z_1^{m_1},\dots,z_n^{m_k}).$$

This follows from results of [28]. Thus all automorphisms of E_m extend smoothly to ∂E_m (see also [2]). Since the automorphism group of B is generated by the unitary group and by the mappings ψ_a (with $m_1 = \dots = m_n = 1$) (see [27], [22]), it follows that any automorphism φ_j of E_m which moves 0 will satisfy $\varphi_j(S) \not\subseteq S$. The genericity of strongly pseudoconvex points in ∂E_m then implies that there is an $s \in S$ such that $\varphi_j(s)$ is strongly pseudoconvex. But then the map φ_j, smooth on the closure of E_m, maps weakly pseudoconvex points to strongly pseudoconvex points and that is impossible. Thus the desired contradiction is obtained and Condition W is not satisfied when $m_1 > 1$.

As a result of the above considerations concerning condition W, theorems of the type which are proved in the present paper can only be true for E_m when $m_1 = 1$. And because of limitations inherent in the proof, only the case $m = (1,1,\dots,1,m_n)$ is considered. It follows (just as was argued in the preceding paragraphs) that if m is of this form and Condition W is satisfied at $P \in \partial E_m$ then P must have the form $(e^{i\theta},0,\dots,0)$. After composing with a rotation, it may be assumed that $P = (1,0,\dots,0) = \mathbb{1}$.

§4. The Three Main Technical Lemmas

The proofs of Theorems 1.1 and 1.2 consist in constructing biholomorphisms as limits of certain normal families of mappings. The work involved is in seeing that the normal limit does not degenerate to a constant mapping. The following simple lemma is the key to seeing that a normal family of mappings does not so degenerate.

<u>Lemma 4.1</u>: Let $\Omega_1, \Omega_2 \subseteq \mathbb{C}^n$ be domains. For $i = 1,2$, let $K_1^i, K_2^i, \dots$ be relatively compact subdomains of Ω_i with the property that if $E \subset\subset \Omega_i$ then there is an $N = N(E)$ such that $K_j^i \supseteq E$ for $j \geq N$. For each j, let f_j be a biholomorphic mapping of a neighborhood U_j^1 of K_j^1 onto $f_j(U_j^1) \subseteq \Omega_2$ such that $f(K_j^1) = K_j^2$.

If there is a $P_0 \in \Omega_1$ and a set $L_0 \subset\subset \Omega_2$ such that $f_j(P_0) \in L_0$ for all j then the sequence $\{f_j\}$ has a subsequence $\{f_{j_k}\}$ with a normal limit mapping $f : \Omega_1 \to \Omega_2$ and f is a biholomorphism.

It should be remarked that variants of this result appear, for instance, in [20] and have been used in [9], [10], [11], [12], [13], [14]. Parts of the proof given in Section 6 may be somewhat novel.

The next Lemma is used only to establish Lemma 4.3. A glance at the statement of Lemma 4.3 reveals that it presupposes the existence of a point $P \in \Omega$ at which $Q_\Omega(P) > 1$. By Bun Wong's Lemma, there is one such point if and only if every $P \in \Omega$ has this property; and this is so if and only if Ω is not biholomorphic to the ball. Therefore, it is essential for the proof of 4.3 to verify

Lemma 4.2: Let Ω be as in the hypotheses of Theorem 1.1 or Theorem 1.2. Then Ω is not biholomorphic to the ball.

Since, by definition, $\partial\Omega$ has $\mathfrak{1}$ as a weakly pseudoconvex point, it is expected that Ω will not be biholomorphic to the ball. But the existing techniques for establihsing biholomoprhic inequivalence of domains requires smooth extension of biholomorphic maps to the boundary (see, for instance, [2]). And the smooth extension results require that $\partial\Omega$ itself be very smooth. since the domain Ω in Theorem 4.1 (resp. 4.2) has only C^3 (resp. C^{n+2}) boundary, it is necessary to devise a new, and non-trivial, method for proving 4.2. See section 7 for the details.

Because of Lemma 4.1, the proofs of Theorems 1.1 and 1.2 amount to constructing a sequence of "approximate biholomorphisms" f_j and proving the non-degeneracy condition consisting of the existence of P_0 and L_0. In fact, what is proved for the situation in Theorems 1.1 and 1.2 is the following result (which should prove useful in other contexts):

Lemma 4.3 (Pseudocollared Neighborhood Lemma): Let Ω be as in Theorem 1.2 and $P_0 \in \Omega$. Then there is a number $\eta_0 = \eta_0(P_0)$, a neighborhood V of $\mathfrak{1}$, and a set

$$S = \{z \in \Omega : z = \zeta + t\nu_\zeta,\ \zeta \in \partial\Omega,\ -\eta\,||\zeta_1|-1|/|\ell n||\zeta_1|-1|| < t < 0\} \subseteq \Omega$$

satisfying the following condition:

(4.3.1) For no $\varphi \in \mathrm{Aut}\,\Omega$ does it hold that $\varphi(P_0) \in S \cap V$.

Lemma 4.2 is used to prove Lemma 4.3. Lemma 4.3 is the key to verifying the hypothesis of Lemma 4.1, and in the proof of Lemma 4.3 lies most of the technical work in this paper. In the next section, assuming the correctness of Lemmas 4.1, 4.2 and 4.3, the proofs of Theorems 1.1 and 1.2 are given.

Lemmas 4.1 and 4.2 are proved in Sections 6 and 7 respectively. The proof of 4.3 is contained in Sections 8 through 13.

§5. The Proofs of Theorems 1.1 and 1.2 (assuming Lemmas 4.1, 4.2, 4.3)

First Theorem 1.1 will be considered.

Fix $m, E_m, \Omega,$ and ρ as in the statement of Theorem 1.1. It may be supposed that there is a $d_0 > 0$ such that

$$\Omega \cap B(\mathbb{1}, 2d_0) \subseteq E_m \cap B(\mathbb{1}, 2d_0). \tag{5.1}$$

From now on, (5.1) is assumed and the subscript μ is omitted.

Let $P_0 = (1-d_0, 0) \in \Omega$ and let S_Ω be the corresponding pseudocollared neighborhood for $\partial\Omega$ and $V = B(\mathbb{1}, 2d_0)$ given by Lemma 4.3. Likewise, let S_{E_m} be the pseudocollared neighborhood for ∂E_m corresponding to P_0. Then

$$S_\Omega = \{z \in \Omega : z = \zeta + t\nu_\zeta,\ \zeta \in \partial\Omega,\ -\eta_1 \| \zeta_1| - 1| / |\ell n| |\zeta_1| - 1\| < t < 0\}$$

$$S_{E_m} = \{z \in E_m : z = \zeta + t\nu_\zeta,\ \zeta \in \partial E_m,\ -\eta_2 \| \zeta_1| - 1| / |\ell n| |\zeta_1| - 1\| < t < 0\}\ .$$

Let η_0 be the minimum of η_1 and η_2. If S_{E_m} is replaced by $S_{E_m} \cap S_\Omega \cap E_m$ and S_Ω is replaced by $S_\Omega \cap S_{E_m} \cap \Omega$ then the conclusion of Lemma 4.3 will be satisfied and it will further hold that

$$\Omega \cap B(\mathbb{1}, 2d_0) \cap S_{E_m} = \Omega \cap B(\mathbb{1}, 2d_0) \cap S_\Omega. \tag{5.2}$$

Replace $\{\varphi_j\}$ by a subsequence if necessary to achieve the condition

$$\varphi_j(P_0) \equiv P_j = (a_j, b_j) \quad \text{with } |1 - a_j| < d_0.$$

Choose $\sigma_j \in \text{Aut } E_m$ such that, setting $h_j = \sigma_j \circ \varphi_j$,

$$h_j(P_0) \equiv \sigma_j \circ \varphi_j(P_0) \equiv P'_j \equiv (a'_j, b'_j)$$

satisfies $a'_j = 1 - d_0$ (refer to Section 3 for details about $\text{Aut } E_m$).

Notice that, by construction, $P_j \notin S_\Omega \cap B(\mathbb{1}, 2d_0)$ hence $P_j \notin S_{E_m} \cap B(\mathbb{1}, 2d_0)$ (line (5.2)). By Lemma 4.3,

$$P'_j \notin S_{E_m} \cap B(\mathbb{1}, 2d_0). \tag{5.3}$$

It follows that

$$P'_j \in E_m \cap \{(1-d_0, z_2) : \operatorname{dist}(z, \partial E_m) \geq 1 - \eta_0(1-d_0)/|\ell n|1-d_0||\}.$$

As a result,

$$L_0 \equiv \{P'_j\} \subset\subset E_m. \tag{5.4}$$

In order to apply Lemma 4.1 with $P_0 = (1-d_0, 0)$ as above and $L_0 = \{P_j\}$, it remains to verify the existence of $\{K_j^1\}$, $\{K_j^2\}$.

Since $T_m : E_m \to B$ is a holomorphic covering and B is complete in the Kobayashi metric, then E_m is also complete. Let

$$K_\ell^2 = \{z \in E_m : \operatorname{dist}_{Kob}^{E_m}(z, 0) < \ell\}.$$

Then

$$K_1^2 \subset\subset K_2^2 \subset\subset \dots \subset\subset E_m$$

and

$$\bigcup_{\ell=1}^{\infty} K_\ell^2 = E_m.$$

If $j = j(\ell)$ is large then

$$\sigma_j^{-1}(K_\ell^2) \subset\subset E_m \cap B(\mathbb{1}, 2d_0)$$

and, necessarily,

$$\sigma_j^{-1}(K_\ell^2) = \{z \in E_m : \operatorname{dist}_{Kob}^{E_m}(z, \varphi_j(P_j)) < \ell\}.$$

By Graham's localization arguments in [8], the existence of a local peaking function at $\mathbb{1}$ now guarantees that if j is sufficiently large then

$$\sigma_j^{-1}(K_\ell^2) \supseteq \{z \in E_m : \operatorname{dist}_{Kob}^{E_m \cap B(\mathbb{1}, 2d_0)}(z, \varphi_j(P_0)) < \ell - 1\}.$$

Since the inclusion map

$$i : \Omega \cap B(\mathbb{1}, 2d_0) \to E_m \cap B(\mathbb{1}, 2d_0)$$

is distance decreasing in the Kobayashi metric, it follows that

$$\sigma_j^{-1}(K_\ell^2) \supseteq \{z \in \Omega : \mathrm{dist}_{\mathrm{Kob}}^{\Omega \cap B(\mathbb{1}, 2d_0)}(z, \varphi_j(P_0)) < \ell - 1\}.$$

Again by the localization of the Kobayashi metric it follows that

$$\sigma_j^{-1}(K_\ell^2) \supseteq \{z \in \Omega : \mathrm{dist}_{\mathrm{Kob}}^{\Omega}(z, \varphi_j(P_0)) < \ell - 2\}.$$

Finally, set

$$K_\ell^1 = \varphi_{j(\ell)}^{-1} \circ \sigma_{j(\ell)}^{-1}(K_\ell^2).$$

Then

$$K_\ell^1 \supseteq \{z \in \Omega : \mathrm{dist}_{\mathrm{Kob}}^{\Omega}(z, P_0) < \ell - 2\}.$$

Clearly

$$\bigcup_{\ell=1}^{\infty} K_\ell^1 = \Omega.$$

Since each $\sigma_{j(\ell)}^{-1}(K_\ell^2)$ is relatively compact in $B(\mathbb{1}, 2d_0) \cap B$ it follows that $K_\ell^1 \subset\subset \Omega$.

Passing to the subsequence $\tilde{h}_j \equiv h_{j(\ell)}$, one sees that all the hypotheses of Lemma 4.1 are satisfied. Thus the mappings $\tilde{h}_j$ have a subsequence with a normal limit h which is a biholomorphism of Ω with E_m.

The proof for $n > 2$ is similar: the transitivity of $\mathrm{Aut}\, E_m$ on normal directions together with the pseudo-collared neighborhood (for tangential directions) prevents degeneration of the normal family.

§6. The Proof of Lemma 4.1

Since Ω_2 is bounded, it is trivial that some subsequence $\{f_{j_k}\}_{k=1}^{\infty} \subseteq \{f_j\}_{j=1}^{\infty}$ has a limit f such that

$$f : \Omega_1 \to \bar{\Omega}_2 .$$

It is necessary to check that in fact $f : \Omega_1 \to \Omega_2$, f is one-to-one, and f is onto.

First notice that $\{f_{j_k}^{-1}\}$ has a subsequence with a normal limit g. Let $Q_0 = f(P_0) \in L \subset\subset \Omega$. Since Ω_1, Ω_2 are hyperbolic, there are positive numbers r_1 and r_2 such that the Kobayashi metric balls $\beta^{\Omega_1}(P_0, r_1)$ and $\beta^{\Omega_2}(Q_0, r_2)$ have compact closures in Ω_1, Ω_2 respectively. Let $r = \min\{r_1, r_2\}$. It follows (by looking at the normal limit of $f_{j_k} \circ f_{j_k}^{-1} = \text{id}$) that

$$f \circ g = \text{id} \quad \text{on} \quad \beta^{\Omega_2}(Q_0, r_2). \tag{4.1.1}$$

Therefore $\det f'$ is not identically zero. Now observe that $\det f'$, being the normal limit of the non-vanishing functions $\det f_j'$, is either identically zero or non-vanishing. Since $\det f'$ is not identically zero, it must be that $\det f'$ is never zero. Hence f is an open mapping and $f(\Omega_1) \subseteq \Omega_2$.

To check the univalence of f, let $B(0,R)$ be a Euclidean ball which contains Ω_1. If $z, w \in \Omega$ are fixed distinct points and if j is so large that $K_j^1 \supseteq \{z,w\}$ then it holds that

$$\begin{aligned} \text{dist}_{\text{Car}}^{K_j^2}(f_j(z), f_j(w)) &= \text{dist}_{\text{Car}}^{K_j^1}(z,w) \\ &\geq \text{dist}_{\text{Car}}^{B(0,R)}(z,w) \equiv \varepsilon_{z,w} > 0. \end{aligned}$$

By the upper semi-continuity of the Cartheodory metric (see [25], [22]), it follows that

$$\text{dist}_{\text{Car}}^{\Omega_2}(f(z), f(w)) \geq \varepsilon_{z,w} > 0.$$

Reversing the roles of f and g yields that $g : \Omega_2 \to \Omega_1$ is one-to-one. Finally, (6.1.1) yields that $f \circ g = \text{id}$ on an open set; hence $f \circ g \equiv \text{id}$ on all of Ω_2. Likewise, $g \circ f \equiv \text{id}$ on Ω_1. This completes the proof. □

§7. The Proof of Lemma 4.2

Since this result is particularly elusive, two proofs will be presented: the first in detail, the second in outine form. The first proof uses localization of the Caratheodary and Eisenman volume forms. The second proof uses an interesting application of Lemma 4.1.

For the first proof, notice that it follows from results of Bell [2], or even from more elementary considerations [4], that E_m is not biholomorphic to the ball. In particular, $Q_{E_m}(0) \equiv 1 + \varepsilon_0 > 1$. Let μ_j be biholomorphic maps of E_m such that $\mu_j(0) = (1-1/j,0,\ldots,0) \longrightarrow \mathbb{1}$. Then of course $Q_{E_m}(\mu_j(0)) = 1 + \varepsilon_0$. Let a_0 be a small positive number. Then the usual localization arguments for M^C and M^E (see [8] or [22], for instance; the delicate quantitative version of localization in Section 11 is not needed here) show that $Q_{E_M \cap B(\mathbb{1},a_0)}(\mu_j(0)) \geq 1 + 3\varepsilon_0/4$ for j large. Now elementary comparisons, using of course the hypothesis (ii) in Theorems 1.1 and 1.2 , shows that $Q_{\Omega \cap B(\mathbb{1},a_0)}(\mu_j(0)) \geq 1 + \varepsilon_0/2$ if j is large. Finally, localization for Q implies that $Q_\Omega(\mu_j(0)) \geq 1 + \varepsilon_0/4$ if j is large. Thus Ω is not biholomorphic to the ball.

A second proof of Lemma 4.2 proceeds as follows. If Ω were biholomorphic to the ball, then it would have a transitive group of biholomorphic self maps. As in the beginning of the proof of Theorem 1.1 in Section 5, replace Ω by

$$\Omega_\mu = \{z : \rho(z_1,(1+\mu)z_2) < 0\}$$

(for a small real μ) so that for some small $d_0 > 0$ it holds that

$$\Omega_\mu \cap B(\mathbb{1},2d_0) \subseteq E_m \cap B(\mathbb{1},2d_0).$$

For j sufficiently large,

$$\varphi_j(P_0) \in \Omega_\mu \cap B(\mathbb{1},2d_0) \subset E_m \cap B(\mathbb{1},2d_0).$$

Write $\varphi_j(P_0) = (a_j,b_j)$. Since $\mathrm{Aut}\,\Omega_\mu$ is transitive, there is an element $\psi_j \in \mathrm{Aut}\,\Omega_\mu$ such that $\psi_j(\varphi_j(P_0)) = (a_j,0) \equiv Q_j \in \Omega_\mu \cap B(\mathbb{1},2d_0)$. Finally, choose $\lambda_j \in \mathrm{Aut}\,E_m$ such that $\lambda_j(Q_j) = 0$.

Then the maps $h_j \equiv \lambda_j \circ \psi_j \circ \varphi_j$ are biholomorphic mappings which satisfy the non-degeneracy conditions of Lemma 4.1 (with $P_0=P_0$ and $L_0=\{0\}$). The existence of K_j^i, U_j^i is established just as in the proof of Theorems 1.1 and 1.2 in Section 5. Therefore the mappings $\{h_j\}$ satisfy the hypotheses of Lemma 4.1 and there is a subsequence converging normally to a biholomorphism of Ω_μ to E_m.

In conclusion, $B \cong \Omega \cong \Omega_\mu \cong E_m$. But it is well-known that $B \ncong E_m$ (see [4]). This contradiction establishes Lemma 4.2.

§8. Uniform Estimates for the $\bar\partial$ Equation

In this section is contained a very detailed form of the uniform estimates for the $\bar\partial$ equation on Ω. The result is essentially due to Beatrous and Range [1]. The contribution here is the more refined estimates on the norm of the right inverse for the $\bar\partial$ operator which is constructed.

Lemma 8.1: Let $0 < n \in \mathbb{Z}$. Then there are positive integers $C_1 = C_1(n)$ and $k_1 = k_1(n)$ such that the following is true.

Let $\Omega \subset\subset \mathbb{C}^n$ be a pseudoconvex domain with C^{N+1} boundary. Let $P \in \partial\Omega$ and let $U_1 \subset\subset U_2 \subset\subset \mathbb{C}^n$ be open neighborhoods of P. Let ρ be a C^{n+1} defining function for Ω with $|\nabla\rho| = 1$ on $\partial\Omega$. Define λ to be the least eigenvalue of the Levi form at points of $\bar{U}_2 \cap \partial\Omega$. Assume that $\lambda > 0$. Let $d = \text{diam}\,(U_1 \cap \Omega)$, $\delta = \text{dist}(U_1, {}^cU_2)$, $D = \text{diam}\,\Omega$. Let $S = \|\rho\|_{C^{n+1}(\mathbb{C}^n)}$. If f is a $\bar\partial$-closed $(0,1)$ form on Ω which has smooth, bounded coefficients which are supported in U_1 then there is a bounded solution u on Ω to the equation $\bar\partial u = f$ with

$$\|u\|_{L^\infty(\Omega)} \leq C_1 \cdot (\mathfrak{X}')^{k_1} \|f\|_{L^\infty(\Omega)}$$

and

$$\|u\|_{L^\infty(\Omega\setminus U_a)} \leq C_1 \cdot (\mathfrak{X}')^{k_1} \|f\|_{L^\infty(\Omega)}$$

where

$$\mathfrak{X}' = n + D + d^{-1} + \lambda^{-1} + \delta^{-1} + S.$$

In what follows, this solution u will be denoted by Tf or $T_\Omega f$. It should be stressed, however, that the operator T depends not only on Ω, but on U_1 and U_2.

Finally, it should be noted that the operator T is linear.

Proof: The main lines of the proof are in [1]. What is new here is the dependence on the parameters.

Now the salient feature of the proof in [1] must be reviewed. First fix an open $U \subseteq \mathbb{C}^n$ such that $U_1 \subset\subset U \subset\subset U_2$ and

$$\begin{aligned}\text{dist}(U_1, {}^cU) &\sim \text{dist}(U, {}^cU_2)\\ &\sim \tfrac{1}{2}\,\text{dist}(U_1, {}^cU_2).\end{aligned}$$

Now, using the "bumping technique" of Kerzman ([17]), one obtains a perturbation $\tilde{\Omega}$ of Ω such that $U \cap \bar{\Omega} \subseteq \tilde{\Omega}$ and the following critical properties hold: Whenever f is a $\bar{\partial}$-closed (0,1) form on Ω with smooth, bounded coefficients supported in U_1, then there are

(i) a $\bar{\partial}$-closed (0,1) form $\tilde{f}$ with smooth, bounded coefficients supported in $\tilde{\Omega} \cap U$ and C^∞ on $\tilde{\Omega}$

and

(ii) a function $u_1 \in C^\infty(\Omega) \cap C(\bar{\Omega})$ with support in U

such that $\tilde{f}$ and u_1 satisfy $f = \tilde{f} + \bar{\partial}u_1$ on Ω and

(a) $\|\tilde{f}\|_{L^\infty(\tilde{\Omega})} \le C_0\|f\|_{L^\infty(\Omega)}$;

(b) $\|u_1\|_{L^\infty(\Omega)} \le C_0\|f\|_{L^\infty(\Omega)}$.

Now $\tilde{f}$, u_1 are constructed using the local solution operator of Kerzman [17] and therefore the constant C_0 is well-known to depend polynomially on $n, D, d^{-1}, \delta^{-1}, \lambda^{-1}, S$.

It is known (see [15]) that local solution operators can be constructed using just two derivatives of the boundary. Kerzman [17] uses just three derivatives, which suffices for the application here.

For the next step, one constructs a small C^{n+1} pseudoconvex perturbation Ω^* of Ω such that $\Omega^*\backslash U_2 = \Omega\backslash U_2$, $\bar{\Omega} \cap U \subseteq \Omega^*$, and $\bar{\Omega}^* \cap U \subseteq \tilde{\Omega}$. The Bochner-Martinelli-Koppelman formula ([21], [22]) is used on $\tilde{\Omega}$ to write $\tilde{f} = g + \bar{\partial}u_2$ where

$$g = \int_{\partial\tilde{\Omega}} \tilde{f} \wedge K_1, \quad u_2 = \int_{\tilde{\Omega}} \tilde{f} \wedge K_0,$$

and K_0, K_1 are the usual Bochner-Martinelli kernels. Since $K_0(z,\cdot)$ is uniformly integrable, one checks that

$$\begin{aligned}\|u_2\|_{L^\infty(\tilde{\Omega})} &\le C_{00}\|\tilde{f}\|_{L^\infty(\tilde{\Omega})} \\ &\le C_{00} \cdot C_0\|f\|_{L^\infty(\Omega)} .\end{aligned}$$

Also the support condition on $\tilde{f}$ makes it straightforward to check that

$u_2 \in C(\bar{\Omega})$. The explicitly known form of K_0 and K_1 make it immediate that C_{00} depends polynomially on $n, D, d^{-1}, \delta^{-1}$.

The support condition on $\tilde{f}$ makes it easy to check (by differentiation under the integral sign) that $g \in C^{n+1}_{(0,1)}(\bar{\Omega}^*)$ and

$$\|g\|_{C^{n+1}_{(0,1)}(\bar{\Omega}^*)} \le C_{000}\|\tilde{f}\|_{L^\infty(\tilde{\Omega})}.$$

As usual, C_{000} depends polynomially on $n, D, d^{-1}, \delta^{-1}$.

Now Kohn's global solution operator developed in [19, Theorem 3.19] provides a map

$$A_{n+1} : C^{n+1}_{(0,1)}(\bar{\Omega}^*) \cap \ker \bar{\partial} \rightarrow C^0(\bar{\Omega}^*) \cap C^\infty(\Omega^*)$$

such that $\bar{\partial}A_{n+1} = \text{id}$ and

$$\|A_{n+1}\alpha\|_{W^{n+1}(\Omega^*,\mu)} \le C'\|\alpha\|_{W^{n+1}_{(0,1)}(\Omega^*,\mu)}.$$

Here μ is a weight function of the form $e^{-t|z|^2}$ and W^{n+1} the corresponding weighted Sobolev class. (Kohn does not state explicitly that W^{n+1} estimates depend on $(n+1)$ derivations of $\partial\Omega$, but this is so and may be checked directly). The choice of weight is fixed once and for all depending on n and $\|\rho\|_{C^{n+1}}$. The weighted Sobolev space is the usual Sobolev space with a different, but comparable norm. The last line, together with the Sobolev Imbedding Theorem ([16, p.123]) now yields that

$$\|A_{n+1}\alpha\|_{L^\infty(\Omega^*)} \le C''\|\alpha\|_{C^{n+1}_{(0,1)}(\Omega^*)}.$$

Now g is $\bar{\partial}$-closed because of the equation $\tilde{f} = g + \bar{\partial}u_2$ so $A_{n+1}g$ makes sense and

$$\begin{aligned}\|A_{n+1}g\|_{L^\infty(\Omega^*)} &\le C''C_{000}\|\tilde{f}\|_{L^\infty(\tilde{\Omega})}\\ &\le C''C_{000}\, C_0\|f\|_{L^\infty(\Omega)}.\end{aligned}$$

Therefore

$$Tf = A_{n+1}g + u_2 + u_1$$

satisfies all the necessary estimates and $\bar{\partial}Tf = f$.

§9. Analysis of Peaking Functions

In a general function algebra, peaking functions play the role that cutoff functions play in the algebra $C^{\infty}(M)$, M a manifold. In the present work, the abstract properties of peaking functions will not suffice and it is necessary to know something about their decay away from the peak point. Further, it is required that the peaking functions continue analytically to a larger domain. This information is contained in the following lemma.

Lemma 9.1: Let $0 < n \in \mathbb{Z}$. Then there are positive integers $C_2 = C_2(n)$ and $k_2 = k_2(n)$ such that the following is true.

Let $\Omega \subset\subset \mathbb{C}^n$ be a pseudoconvex domain with C^{n+1} boundary. Let $P \in \partial\Omega$ and let $U_1 \subset\subset U_2 \subseteq \mathbb{C}^n$ be open neighborhoods of P. Let ρ be a C^{n+1} defining function for Ω with $|\nabla\rho| = 1$ on $\partial\Omega$. Define λ to be the least eigenvalue of the Levi form at points of $\bar{U}_2 \cap \partial\Omega$. Assume that $\lambda > 0$. Let $d = \operatorname{diam}(U_1 \cap \Omega)$, $\delta = \operatorname{dist}(U_1, {}^cU_2)$, $D = \operatorname{diam} \Omega$, $S = \|\rho\|_{C^{n+1}(\mathbb{C}^n)}$. Then there is a perturbation Ω^* of Ω such that $\Omega^* \setminus U_2 = \Omega \setminus U_2$ and $\bar{\Omega} \cap U_1 \subseteq \Omega^*$ and there is a function ψ_P holomorphic on Ω^* such that

(9.1.1) $\psi_P(P) = 1$

(9.1.2) $1/2 \le |\psi_P(z)| < 1$ for $z \in \Omega \cup (\partial\Omega \cap U_1)\setminus\{P\}$

(9.1.3) $|\psi_P(z)| \le 1 - \omega \cdot \beta(z,P)$

where

$$C_2(\mathfrak{X}')^{-k_2} \le \omega \le C_2(\mathfrak{X}')^{k_2}$$

and $\mathfrak{X}' = n + \delta^{-1} + d^{-1} + \lambda^{-1} + D + S$ and β is the skew distance defined at the end of Section 2.

Proof: This is the standard construction of a peaking function at a strongly pseudoconvex point which can be found in [22]. The function is constructed locally as the exponentiated Levi polynomial. The estimates (9.1.1)-(9.1.3) then hold locally by inspection. The peaking function is then defined globally by using a cutoff function and solving a suitable $\bar{\partial}$ problem. The derivatives of the cutoff function introduce polynomial dependence on δ^{-1}, and the Levi polynomial and estimates for the $\bar{\partial}$ problem in Lemma 8.1 introduce polynomial dependence on the other parameters.

Finally the resulting peaking function φ is replaced by $\psi \equiv (\varphi+3)/4$ to guarantee that (9.1.2) holds globally. □

§10. Uniform Approximation of Holomorphic Functions

<u>Proposition 10.1</u>: Let $\Omega \subset\subset \mathbb{C}^n$ be pseudoconvex with C^{n+1} boundary. Let $P \in \partial\Omega$ and let $U_1 \subset\subset U_2 \subset\subset \mathbb{C}^n$ be neighborhoods of P. Assume that the eigenvalues of the Levi form on $\partial\Omega$ are bounded from 0 by $\lambda > 0$ on $\bar{U}_2 \cap \partial\Omega$. Let ψ_P be the peaking functin given by Lemma 9.1. Suppose that there are numbers $0 < a < b < 1$ such that

$$U_2 = \{z \in \mathbb{C}^n : |\psi_P(z)| > a\}, \quad U_1 = \{z \in \mathbb{C}^n : |\psi_P(z)| > b\}.$$

Let S, d, δ, D be as in Lemmas 8.1, 9.1 and let $\varepsilon > 0$ and $0 \leq k \in \mathbb{Z}$. Then there is a constant L such that if f is <u>any</u> bounded holomorphic functions on $U_2 \cap \Omega$ then there is a bounded holomorphic $\hat{f}$ on Ω such that

$$\text{(10.1.1)} \qquad \sup_{z \in U_1 \cap \Omega} \left| \left(\frac{\partial}{\partial z}\right)^{\alpha} f(z) - \left(\frac{\partial}{\partial z}\right)^{\alpha} \hat{f}(z) \right| \leq \varepsilon \|f\|_{L^\infty(U_2 \cap \Omega)},$$

for all multi-indices α such that $|\alpha| \leq k$, and

$$\text{(10.1.2)} \qquad \|\hat{f}\|_{L^\infty(\Omega)} \leq L \cdot \|f\|_{L^\infty(U_2)}.$$

Here

$$L \leq \{(C_3 \mathfrak{X}')^{k_1}/(\varepsilon c^k)\}^{4/(b-a)}$$

where $\mathfrak{X}' = n + d^{-1} + \delta^{-1} + \lambda^{-1} + D + s$ and C_3, c, k_1 are positive constants.

<u>Proof</u>: This follows the ideas in [8, p.230]. It will be supposed, without loss of generality, that $b > a > 3/4$. Set $a' = (3a+b)/4$, $b' = (a+3b)/4$. Let $U_2' = \{z \in \Omega : |\psi_P(z)| > a'\}$, $U_1' = \{z \in \text{dom}\, \psi_P : |\psi_P(z)| > b'\}$. Select $\tilde{\Omega}$ so that

$$\text{(i)} \quad U_1 \subset\subset \tilde{\Omega}$$

$$\text{(ii)} \quad \tilde{\Omega} \setminus U_1' = \Omega \setminus U_1'$$

$$\text{(iii)} \quad |\varphi_P| \geq b' \quad \text{on} \quad \tilde{\Omega}\setminus\Omega.$$

Let $\tilde{U}_1 = (\tilde{\Omega}\setminus\Omega) \cup (U_1'\cap\Omega)$. Define a cutoff function $\eta \in C^\infty(\Omega)$ such that $0 \leq \eta \leq 1$, $\eta = 1$ on $\Omega \cap U_2'$, $\eta = 0$ on $\Omega\setminus U_2$, $|\nabla\eta| \leq \tilde{C}_3\delta^{-1}$, where $\delta = \text{dist}(U_1, {}^cU_2)$. Choose $c > 0$ such that

$$\text{dist}(U_1, \Omega\setminus U_1') \geq c\delta,$$

$$\text{dist}(U_1, \partial\tilde{U}_1) \geq c\delta .$$

Notice that c, $\tilde{C}_3$ may be chosen to depend polynomially on the usual parameters.

Now given $f \in H^\infty(U_2\cap U)$, define

$$g(z) = \begin{cases} (\bar{\partial}\eta(z))f(z) & \text{if } z \in U_2 \cap \tilde{\Omega} \\ 0 & \text{if } z \in \tilde{\Omega}\setminus U_2. \end{cases}$$

Then $g \in L^\infty_{(0,1)}(\tilde{\Omega})$, g is smooth, and $\bar{\partial}g = 0$. Notice that $g = 0$ on U_2'; hence $|\psi| \leq a'$ on supp g. Now let

$$\hat{f} = \eta \cdot f - \psi_P^{-r} \; T_{\tilde{\Omega}}(\psi_P^r g)$$

where $T_{\tilde{\Omega}}$ is the $\bar{\partial}$ solution operator of Lemma 8.1 and $0 < r \in \mathbb{Z}$ will be selected.

Direct calculation shows that $\hat{f}$ is holomorphic and Lemma 8.1 concerning the operator $T_{\tilde{\Omega}}$ shows that $\hat{f} \in H^\infty(\tilde{\Omega})$. Furthermore,

$$\begin{aligned} \|\hat{f}-f\|_{L^\infty(\tilde{U}_1)} &= \|\psi^{-r}T_{\tilde{\Omega}}(\psi^r g)\|_{L^\infty(\tilde{U}_1)} \\ &\leq C_1(\tilde{\mathfrak{X}}')^{k_1}(b')^{-r}\|\psi^r g\|_{L^\infty(\tilde{\Omega})} \\ &\leq C_1(\tilde{\mathfrak{X}}')^{k_1}(a'/b')^{-r}\|g\|_{L^\infty(\tilde{\Omega})} \\ &\leq C_1\tilde{C}_3\delta^{-1}(\tilde{\mathfrak{X}}')^{k_1}(a'/b')^{-r}\|f\|_{L^\infty(U_2\cap\Omega)}. \end{aligned}$$

Here it is understood that $\tilde{\mathfrak{X}}'$ is the constant $\mathfrak{X}'$ from Lemma 8.1 computed relative to $\tilde{\Omega}$, $\tilde{U}_1$, U_2'. Now let

$$r = \left[\ln \left[\frac{k!C_1(\tilde{\mathfrak{X}}')^{k_1}\tilde{C}_3}{\delta^k \varepsilon c^k} \right] / \ln(b'/a') \right] + 1$$

where [ ] is the greatest integer function. Then the Cauchy estimates apply to the holomorphic function $\hat{f} - f = \psi^{-r}S(\psi^r f)$ on $\tilde{U}_1$. For $z \in U_1 \cap \Omega \subset\subset \tilde{U}_1$ and $|\alpha| = k$ one thus obtains

$$\begin{aligned} \left\| \left(\frac{\partial}{\partial z}\right)^{\alpha} (\hat{f}-f) \right\|_{L^\infty(U_1 \cap \Omega)} &\le k!(c\delta)^{-k} \|\hat{f}-f\|_{L^\infty(\tilde{U}_1)} \\ &\le k!(c\delta)^{-k} C_1(\mathfrak{X}')^k C_3 \delta^{-1}(a'/b')^r \|f\|_{L^\infty(U_2 \cap \Omega)} \\ &< \varepsilon \|f\|_{L^\infty(U_2 \cap \Omega)}, \end{aligned}$$

by the choice of r. Also

$$\begin{aligned} \|\hat{f}\|_{L^\infty(\Omega)} &\le \|f\|_{L^\infty(U_2 \cap \Omega)} + \|\psi^{-r} T_{\tilde{\Omega}}(\psi^r g)\|_{L^\infty(\Omega)} \\ &\le \|f\|_{L^\infty(U_2 \cap \Omega)} + 2^r \|T_{\tilde{\Omega}}(\psi^r g)\|_{L^\infty(\Omega)} \\ &\le \|f\|_{L^\infty(U_2 \cap \Omega)} + 2^r C_1(\tilde{\mathfrak{X}}')^{k_1} \|\psi^r g\|_{L^\infty(\Omega)} \\ &\le \|f\|_{L^\infty(U_2 \cap \Omega)} + 2^r C_1(\tilde{\mathfrak{X}}')^{k_1} (a')^r \tilde{C}_3 \delta^{-1} \|f\|_{L^\infty(U_2 \cap \Omega)}. \end{aligned}$$

This last is obtained by recalling the definition of g, the support of g, and the estimate on $|\nabla \eta|$. This is

$$\le \left\{ 1 + \left[\frac{k!C_1(\tilde{\mathfrak{X}}')^{k_1}\tilde{C}_3}{\delta^{k+1} \varepsilon c^k} \right]^{\frac{2a'}{b'-a'}} \frac{\varepsilon \delta^k c^k}{m!} \right\} \|f\|_{L^\infty(U_2 \cap \Omega)},$$

by the definition of r,

$$\leq \left\{ 1 + \left[\frac{k! C_1(\tilde{\mathfrak{X}}')^{k_1} \tilde{C}_3}{\delta^{k+1} \varepsilon c^k} \right]^{\frac{4}{b-a}} \right\} \|f\|_{L^\infty(U_2 \cap \Omega)},$$

$$\equiv L\|f\|_{L^\infty(U_2 \cap \Omega)}$$ □

§11. Localization of the Caratheodary and Eisenman Volume Forms

In order to use Q_Ω in later normal families arguments, it is necessary now to compute the dependence of M_Ω^C and M_Ω^E on local differential boundary data. This computation necessitates explicit formulation of certain localization arguments.

The localization arguments for the metrics under consideration here were pioneered by Graham in [8]. His work was done in a C^2 strictly pseudoconvex domain, and his localization estimates are therefore uniform over points of the boundary. By contrast it is necessary in the present paper to localize the metrics at strongly pseudoconvex points which are very near to a weakly pseudoconvex point: as the Levi form degenerates, how do the estimates change?

The estimates obtained below should have broad applicability to the function theory on domains in which strongly pseudoconvex boundary points are generic. The thrust of the calculations is to see that no global information about the domain except for its diameter plays a role in the localization arguments.

<u>Proposition 11.1</u>: Let $\Omega \subset\subset \mathbb{C}^n$ be pseudoconvex with C^{n+1} boundary. Let $P \in \partial\Omega$ and suppose that $U_2 \subseteq \mathbb{C}^n$ is a neighborhood of P such that the eigenvalues of the Levi form on $\partial\Omega$ are bounded from 0 by $\lambda > 0$ on $\bar{U}_2 \cap \partial\Omega$. Suppose further that U_2 has the form

$$U_2 = \{z : |\psi_P(z)| > a\}$$

where ψ_P is the peaking function of Lemma 9.1. Let $\varepsilon > 0$. Then there is a constant b', $a < b' < 1$, such that if $U_1' = \{z : |\psi_P(z)| > b'\}$ and if $w \in U_1'$ then

$$1 \leq \frac{M_{\Omega \cap U_2}^C(w)}{M_\Omega^C(w)} \leq 1 + \varepsilon .$$

Here

$$b' = 1 - c' \cdot \varepsilon \cdot (1-a)^2/(\ln \mathfrak{X}),$$

c' is a small positive constant, and $\mathfrak{X} = n + D + \varepsilon^{-1} + \lambda^{-1} + S + (1-a)^{-1}$.

Proof: Let $\eta > 0$ be such that $\frac{(1+\eta)^n}{(1-\eta)^n} = 1 + \varepsilon$. Let $w \in U_1'$. Select a function $F \in B(\Omega \cap U_2)$ such that $F(w) = 0$ and $|\det F'(w)| = M^C_{\Omega \cap U_2}(w)$ (this can be done since B is taut -- see [22]). Then $F = (F_1, \ldots, F_n)$ is an n-tuple of complex valued, bounded functions.

Set $b = (a+2)/3$ and $U_1 = \{z : |\psi_p(z)| > b\}$. The approximation result 10.1 may thus be applied to Ω, U_1, U_2 and each F_j with $m=1$ and ε replaced by η/n. One therefore obtains a function $\hat{F} : \Omega \to \mathbb{C}^n$ such that

$$\|\hat{F}-F\|_{L^\infty(U_1 \cap \Omega)} < \eta$$

$$\|\hat{F}\|_{L^\infty(\Omega)} \le L$$

$$\hat{F}(w) = F(w) = 0$$

$$\hat{F}'(w) = F'(w).$$

(The last two conditions are arranged, of course, by subtracting off a linear polynomial with small coefficients. The constant L is specified in 10.1.) Set

$$\beta = \frac{60}{(1-a)^2} \ln \left[\frac{10\, C_1 (\tilde{\tilde{\mathfrak{X}}}')^{k_1}}{c\delta^2 \varepsilon} \right]$$

where $\tilde{\tilde{\mathfrak{X}}}'$ is the constant $\mathfrak{X}'$ from Lemma 10.1 computed relative to Ω and the sets

$$U_1 = \{z : |\psi_p(z)| > (a+2)/3\}$$

and

$$U_2 = \{z : |\psi_p(z)| > a\}.$$

Define

$$\tilde{F}(z) = (\psi_p(z))^\beta \cdot \hat{F}(z).$$

Notice that

$$\tilde{F}(w) = \hat{F}(w) = 0$$

and

$$\tilde{F}'(w) = \psi_P^\beta(w)\hat{F}'(w)$$
$$= \psi_P^\beta(w)F'(w).$$

Therefore

$$|\det \tilde{F}'(w)| \geq |\psi_P^\beta(w)|^n \cdot |\det F'(w)|$$
$$= |\psi_P^\beta(w)|^n \, M_{U_2 \cap \Omega}^C(w)$$

by the choice of F. Notice that since $w \in U_1'$ it holds that $|\psi(w)| \geq b'$ so

$$|\psi_P^\beta(w)| \geq b'^\beta.$$

The choice of b' and β and a little calculation then reveal that

$$|\psi_P^\beta(w)| \geq (1-\eta).$$

Thus

$$|\det \tilde{F}'(w)| \geq (1-\eta)^n M_{U_2 \cap \Omega}^C(w).$$

On the other hand, by the approximation result,

$$\|\tilde{F}\|_{L^\infty(\Omega \setminus U_1)} \leq b^\beta L \leq 1.$$

The value of L is given in the proof of 10.1 and the estimate follows by direct calculation. Also

$$\|\tilde{F}\|_{L^\infty(U_1 \cap \Omega)} \leq \|F\|_{L^\infty(U_1 \cap \Omega)} + \eta$$
$$\leq 1 + \eta.$$

So $\frac{1}{1+\eta} \cdot \tilde{F} \equiv G \in B(\Omega)$ and

$$M_\Omega^C(w) \geq |\det G'(w)| \geq \frac{(1-\eta)^n}{(1+\eta)^n} \, M_{\Omega \cap U_2}^C(w)$$
$$= \frac{1}{(1+\varepsilon)} \cdot M_{\Omega \cap U_2}^C(w). \qquad \Box$$

<u>Proposition 11.2</u>: Let hypotheses be as in Proposition 11.1. Let $\varepsilon > 0$. Then there is a b'', $a < b'' < 1$, such that if $U_1'' = \{z : |\psi_P(z)| > b''\}$ and $w \in U_1''$ then

$$1 \leq \frac{M^E_{U_2 \cap \Omega}(w)}{M^E_\Omega(w)} < 1 + \varepsilon .$$

Indeed, the choice

$$b'' = 1 - c''\varepsilon^{4n}(1-a)^{4n}$$

will do, c'' a small positive constant.

<u>Proof</u>: Fix $w \in \Omega \cap U_1''$. Let $F \in \Omega(B)$, $F(0) = w$. Consider $\psi_P \circ F$. Notice that

$$b'' \leq |\psi_P \circ F(0)| \leq \frac{1}{\Lambda_n} \int_B |\psi_P \circ F(\zeta)| d\, \mathrm{vol}(\zeta),$$

where Λ_n is the Euclidean volume of the ball $B \subseteq \mathbb{C}^n$. Set $\hat{b}'' = 1-(c'')^{1/2}\varepsilon^{2n}(1-a)^{2n}$. Since $\sup|\psi_P \circ F| \leq 1$, it follows that

$$\mathrm{vol}\{z \in B : |\psi_P \circ F| > \hat{b}''\} \geq \hat{b}'' \cdot \Lambda_n .$$

Set

$$W = \{\zeta \in \Omega : |\psi_P(\zeta)| > b''\}$$

and

$$E = \{z \in B : F(z) \in W\}.$$

Then $0 \in E$ and $\mathrm{vol}(E) \geq \hat{b}''\Lambda_n$. Let $\tau = \varepsilon/2n$. Notice that every point of $B(0,1-\tau)$ is within Euclidean distance $(1-\hat{b}'')^{1/2n}$ of an element of E (since the volume of E is so large). Thus, for $v \in B(0,1-\tau)$,

$$|\psi_P(v)| \geq |\psi_P(e)| - |\psi_P(e)-\psi_P(v)|$$

(where $e \in E$ is an element satisfying $|e-v| \leq (1-\hat{b}'')^{1/2n}$) and this last is, by the Cauchy estimates,

$$\geq b'' - \frac{2(1-\hat{b}'')^{1/2n}}{\tau} \equiv b'''.$$

Consider the function

$$\tilde{F} : B \longrightarrow \Omega \cap \{z : |\psi_P(z)| > b'''\} \equiv \Omega \cap U_1'''$$
$$z \longrightarrow F((1-\tau)z).$$

Notice that $\tilde{F}(0) = F(0) = w$. It follows that

$$M^E_{\Omega \cap U_1'''}(w) \le \frac{1}{|\det \tilde{F}'(0)|} = \frac{1}{|(1-\tau)^n|} \frac{1}{|\det F'(0)|} .$$

Since this inequality holds for any $F \in \Omega(B)$ such that $F(0) = w$, it follows that

$$M^k_{\Omega \cap U_1''}(w) \le \frac{1}{|(1-\tau)^n|} \cdot M^E_\Omega(w)$$
$$< (1+\varepsilon) M^E_\Omega(w).$$

It remains to check that

$$M^E_{\Omega \cap U_2}(w) \le M^E_{\Omega \cap U_1''}(w).$$

For this, it is enough to see that

$$\Omega \cap U_1''' \subseteq \Omega \cap U_2$$

or that $b''' \ge a$. This last assertion follows by an easy calculation. □

<u>Corollary 11.3</u>: Let Ω, P be as in the previous two propositions. Let ψ_P be the peaking function at P which is provided by Proposition 9.1. Suppose that

$$U_2 = \{z : |\psi_P(z)| > a\}$$

is a neighborhood of P such that the eigenvalues of the Levi forms on $\partial\Omega \cap \bar{U}_2$ are bounded from 0 by $\lambda > 0$. Let $\varepsilon > 0$ and let

$$b_0 = 1 - C_4 \cdot \varepsilon^{4n}(1-a)^{4n}/(\ln \mathfrak{X})$$

(with $\mathfrak{X} = n + D + \varepsilon^{-1} + \lambda^{-1} + S + (1-a)^{-1}$ and C_4 a suitable constant). Define

$$U_0 = \{z \in \mathbb{C}^n : |\psi_P(z)| > b_0\}.$$

If $w \in U_0 \cap \Omega$ then

$$(1-\varepsilon)Q_\Omega(w) \le Q_{\Omega \cap U_2}(w) \le (1+\varepsilon)Q_\Omega(w).$$

Proof: This is immediate from the two propositions. □

It is the Corollary which will be used in what follows to estimate $Q_\Omega(w)$.

§12. Estimates on the Biholomorphic Invariant Q_Ω

The purpose of this section is to prove the crucial estimates on Q_Ω which will lead to Lemma 4.3. First a technical lemma is needed.

Lemma 12.1: Let $Q : \mathbb{C}^n \times \mathbb{C}^n \to \mathbb{C}$ be a positive definite quadratic form. Let

$$\mathcal{D}_Q = \{z \in \mathbb{C}^n : \text{Re } z_1 < -Q(z,z)\}.$$

Then there is a biholomorphic map

$$\Phi_Q : \mathcal{D}_Q \to B.$$

Proof: (see [8]). Write $Q(w,w) = a_{ij}w_i\bar{w}_j$. Since Q is positive definite, one can diagonalize $\sum_{i,j=2}^{n} a_{ij}w_i\bar{w}_j$. So $\mathcal{D}_Q$ is biholomorphic to a region of the form

$$\text{Re } w_1 < -\sum_{j=2}^{n} |\alpha_j w_j|^2 - 2\,\text{Re} \sum_{j=2}^{n} a_{j1}w_j\bar{w}_1 - a_{11}|w_1|^2.$$

A change of variable of the form

$$z_1 = w_1$$
$$z_j = w_j + \beta_j w_1, \quad j = 2,\ldots,n,$$

eliminates the cross terms and yields

$$|\alpha_0(z_1-\beta_1)|^2 + \sum_{j=2}^{n} |\alpha_j z_j|^2 < \gamma.$$

Finally, translation and dilation of z_1, together with dilations of $z_2,\ldots,z_n$, give the region

$$\sum_{j=1}^{n} |z_j|^2 < 1. \qquad \square$$

Notice that the biholomorphism of $\mathcal{D}_Q$ to B is not unique, but applications of this lemma in what follows will, for convenience, use the particular biholomorphism Φ_Q constructed in the proof.

Proposition 12.2: Let $\Omega \subset\subset \mathbb{C}^n$ be pseudoconvex with C^{n+1} boundary. Let $P \in \partial\Omega$ be a strongly pseudoconvex point and ψ_P the associated peaking function. Let $U_2 = \{z : |\psi_P(z)| > a\}$ be a neighborhood of P such that the Levi form of $\partial\Omega$ has eignevalues bounded from 0 by $\lambda > 0$ on $\partial\Omega \cap \bar{U}_2$. Let $\varepsilon > 0$. There is a number b_0, $a < b_0 < 1$, such that if

$$U_0 = \{z : \psi_P(z) > b_0\}$$

and if $w \in U_0 \cap \Omega$ and the orthogonal projection of w to $\partial\Omega$ is P then

$$1 < Q_\Omega(w) < 1 + \varepsilon.$$

The number b_0 may be taken to be

$$b_0 = 1 - C_5(1-a)^{4n}\varepsilon^{4n}/(\ln(\mathfrak{X}))$$

where

$$\mathfrak{X} = n + D + \varepsilon^{-1} + \lambda^{-1} + S + (1-a)^{-1}.$$

Proof: By a standard calculation (see [22, Lemma 3.23]), there is a neighborhood W of P and a local change of coordinates on W such that P becomes the origin and

$$\rho(z) = \operatorname{Re} z_1 + L_P(z) + N(z)$$

is a defining function for Ω in W. Here the quadratic form L_P is the Levi form and $N(z) = O(|z|^3)$. Suppose in advance that $U_2 \subseteq W$, and from now on work in the new coordinates. Set $\eta = \varepsilon\lambda/10$ and define

$$E^- = E_\eta^-(P) = \{z : \operatorname{Re} z_1 + L_P(z) + \eta|z|^2 < 0\},$$
$$E^+ = E_\eta^+(P) = \{z : \operatorname{Re} z_1 + L_P(z) - \eta|z|^2 < 0\}.$$

There are constants $\mu_2 < \mu_1 < 1$ such that if

$$V_1 = \{z : |\psi_P(z)| > \mu_1\},$$
$$V_2 = \{z : |\psi_P(z)| > \mu_2\},$$

then

$$E^- \cap V_1 \subseteq \Omega \cap V_2 \subseteq E^+.$$

Of course the choice of μ_1, μ_2 depend on ψ_P and polynomially on $\lambda^{-1}, \|\rho\|_{C^2}$.

According to Corollary 11.3, there is a number $b_0, \mu_1 < b_0 < 1$, such that if $w \in U_0 \equiv \{z : |\psi_P(z)| > b_0\}$ and $w \in \Omega$ then

$$(1-\varepsilon/3)Q_\Omega(w) \le Q_{\Omega\cap V_1}(w) \le (1+\varepsilon/3)Q_\Omega(w).$$

A check of the form of the constant b_0 shows that, with a negligible adjustment in b_0, it also holds that

$$\left(1 - \frac{\varepsilon}{3}\right)M_{E^-}(w) \le M_{E^-\cap V_1}(w) \le \left(1 + \frac{\varepsilon}{3}\right)M_{E^-}(w)$$

and

$$\left(1 - \frac{\varepsilon}{3}\right)M_{E^+}(w) \le M_{E^+\cap V_1}(w) \le \left(1 + \frac{\varepsilon}{3}\right)M_{E^+}(w).$$

Here M can be taken to be M^C or M^{E^*}. Now for $w \in U_0$ it holds that

$$\begin{aligned}
1 \le Q_\Omega(w) &= \frac{M^E_\Omega(w)}{M^C_\Omega(w)} \\
&\le \frac{M^E_{\Omega\cap V_2}(w)}{M^C_{\Omega\cap V_1}(w)\cdot\left(1 - \frac{\varepsilon}{3}\right)} \\
&\le \frac{M^E_{E^-\cap V_1}(w)}{M^C_{E^+\cap V_1}(w)\cdot\left(1 - \frac{\varepsilon}{3}\right)} \\
&\le \frac{M^E_{E^-}(w)\cdot\left(1 + \frac{\varepsilon}{3}\right)}{M^C_{E^+}(w)\left(1 - \frac{\varepsilon}{3}\right)^2}
\end{aligned}$$

$$= \frac{M^{E}_{E^-}(w)\cdot\left(1+\frac{\varepsilon}{3}\right)}{|\mathcal{J}|\cdot M^{C}_{E^-}(w)\cdot\left(1-\frac{\varepsilon}{3}\right)^2}$$

where $\mathcal{J}$ is the Jacobian determinant of the biholomorphism $\Phi : E^+ \to E^-$ (provided by Lemma 12.1) evaluated at w. An easy calculation shows that $|\mathcal{J}| \geq 1 - c\varepsilon$, some $c > 0$. So the last line is majorized by

$$\frac{\left(1+\frac{\varepsilon}{3}\right)}{(1-c\varepsilon)\cdot\left(1-\frac{\varepsilon}{3}\right)^2} .$$

Replacing ε by $\varepsilon/(3(c+1))$ completes the proof. □

§13. Proof of Lemma 4.3.

By Lemma 4.2, Ω is not holomorphic to the ball. So there is a $P_0 \in \Omega$ such that $Q_\Omega(P_0) \equiv 1 + 2\varepsilon > 0$. Let

$$C_\varepsilon = \{z \in \Omega : Q_\Omega(z) < 1 + 2\varepsilon\}.$$

Then $\varphi(C_\varepsilon) = C_\varepsilon$ for every $\varphi \in \operatorname{Aut}\Omega$. Likewise $\varphi(P_0) \notin C_\varepsilon$ for every $\varphi \in \operatorname{Aut}\Omega$. The Lemma will be established if it can be shown that C_ε contains a pseudo-collared neighborhood as described in the statement of 4.3.

For simplicity the proof is given in dimension two. The case of $\Omega \subseteq \mathbb{C}^n$, $n > 2$, is just the same. The proof begins by analyzing Q_Ω in the eggs E_m.

Fix a positive integer m; set

$$E_m = \{(z_1, z_2) : |z_1|^2 + |z_2|^{2m} < 1\},$$

and

$$\partial E_m \ni P = (1-\delta_1, \delta_2)$$

where $\delta_1, \delta_2 \in \mathbb{C}$ are small and non-zero with $\operatorname{Re}\delta_1 > 0$. Let ν_P be the unit outward normal at P and $\varepsilon > 0$. For which values of t is it true that

$$Q_{E_m}(P - t\nu_P) < 1 + \varepsilon?$$

To answer this question, let

$$W_2^P = \left\{ (z_1, z_2) \; : \; |z_1 - (1-\delta_1)| < \frac{|\delta_1|}{2} \; , \; |z_2 - \delta_2| < \frac{|\delta_3|}{2} \right\} .$$

Notice that for $z \in W_P^2 \cap \partial E_m$ the eigenvalues of the Levi form at z are bounded from 0 by $c \cdot |\delta_2|^{2m-2}$, c is a small positive constant.

Introduce new coordinates centered at P by taking w_1 to be the complex line generated by ν_P and w_2 the orthogonal complex line. Orient w_1 so that $(1+i0, 0+i0)$ points in the direction ν_P. Let

$$\Psi : E_m \longrightarrow F_m$$

be given by

$$\Psi(w_1, w_2) = (w_1/|\delta_2|^{2m}, w_2/|\delta_2|),$$

where F_m is <u>defined</u> to be $\Psi(E_m)$. Direct calculation then shows that the eigenvalues of the Levi form on $\Psi(U_2^P) \cap \partial F_m$ are bounded from 0 by some constant c' which does not depend on δ_1, δ_2.

To apply 11.3 to F_m at $\Psi(P)$ one may take

$$\begin{aligned} \psi_{\Psi(P)}(v_1, v_2) &= \exp v_1, \\ a &= \frac{9}{10} \\ d &= 1/2 \\ \delta &= \frac{1}{100} \\ D &= C/|\delta_2|^{2m}. \\ S &= C' \end{aligned}$$

where C, C' are constants independent of δ_1, δ_2. Then 10.3 says that

$$Q_{F_m}(\Psi(P - t\nu_P)) < 1 + \varepsilon$$

provided that

$$\Psi(P - t\nu_P) \in \{v : |\psi_{\Psi(P)}(v)| > b_0\}$$

where b_0 may be taken to be

$$b_0 = 1 - C''\varepsilon^{4n}/|\ln|\delta_2| \ln \varepsilon| .$$

Write

$$\Psi(P-t\nu_P) = P' - t'\nu_{P'} .$$

The definition of Ψ_P shows that if

$$0 < t' < C''' \varepsilon/|\ln|\delta_2|\ln \varepsilon|$$

then

$$Q_{F_m}(\Psi(P'-t'\nu_{P'})) < 1 + \varepsilon .$$

Now Q is invariant under the biholomorphic map Ψ so it follows that if

$$0 < t < C'''|\delta_1|\varepsilon^{4n}/|\ell n|\delta_2|\ell n \varepsilon|$$

then

$$Q_{E_m}(P-t\nu_P) < 1 + \varepsilon .$$

Now if $\Omega \subset\subset \mathbb{C}^2$ is a domain satisyfing the hypotheses of Theorem 1.1 then Ω is pseudoconvex. This is true because if $a_0 > 0$ is small then $B(\mathbf{1},a_0) \cap \Omega$ is pseudoconvex. Exhaust $B(\mathbf{1},a_0) \cap \Omega$ by an increasing sequence of smooth strictly pseudoconvex subdomains $U_1 \subset\subset U_2 \subset\subset \dots$. Then the sets $\varphi_1^{-1}(U_{\ell(1)}), \varphi_2^{-1}(U_{\ell(2)}),\dots$ form an increasing exhaustion of Ω by smooth strictly pseudoconvex domains, provided that the subsequence $\ell(j)$ of $\mathbb{N}$ increases rapidly. Thus Ω is pseudoconvex.

Then all of the analysis, that is the localization and change of scale, which has just been performed on E_m applies on Ω at points $P = (1-\delta_2,\delta_2) \in \partial\Omega$ near $\mathbb{1}$ (of course ψ_P from Section 9 must be used instead of the explicit peaking function which can be defined on the egg). It follows that if

$$0 < t < C_0|\delta_1|\varepsilon^{4n}/|\ell n|\delta_1|\ell n \varepsilon|$$

then

$$Q_\Omega(P-t\nu_P) < 1 + \varepsilon .$$

Since $\ell n|\delta_2| \sim \ell n|\delta_1|$, we are done.

§14. Concluding Remarks and Speculations

The techniques of this paper are very special in the sense that they require Ω to have a nearly transitive group action in the complex normal direction at the distinguished boundary point (1,0). The pseudocollared neighborhood (Lemma 4.3) prevents degeneracy of the normal family of maps in the (remaining) tangential directions. While Lemma 4.3 is true in considerably greater generality, it is of much less use in the absence of control over the normal directions.

It would be very attractive to have a version of our theorem in which the model domain is any domain of finite type. In the case of finite type, the action of Aut Ω on Ω extends to an action of Aut Ω on $\partial\Omega$. And restrictions on this action could in principle be calculated from considerations of boundary invariants of CR maps. Of course the Chern-Moser-Tanaka invariant theory has been developed for strongly pseudoconvex domains only; a detailed theory for weakly pseudoconvex domains, even those of finite type, does not yet exist.

A by product of the successful completion of the program indicated in the preceding paragraph might be an understanding of domains with compact automorphism group. That is, if Ω does not possess enough automorphisms to carry an interior point P_0 arbitrarily near a boundary point Q_0, and if the obstruction is local, then it should be characterized by a specific finite order jet of the defining function. The normal family techniques of the present paper shed no light on this problem.

Finally, R. Remmert has suggested that, in the case of non-compact Aut Ω, we attempt to relate the rank of the Levi form at the limit point Q_0 to the dimension of Aut Ω. This will be a subject for future work.

References

1. F. Beatrous and R.M. Range, On holomorphic approximation in weakly pseudoconvex domains, Pac. Jour. Math. 89(1980), 249-255.
2. S.R. Bell, Biholomorphic mappings and the $\bar{\partial}$ problem, Annals of Math. 114(1981), 103-113.
3. T. Bloom and I. Graham, A geometric characterization of points of finite type, J. Diff. Geom. 12(1977), 171-182.
4. R. Braun, W. Karp and H. Upmeier, On the automorphisms of circular domains in complex Banach spaces, Manuscripta Math. 25(1978), 97-133.
5. D. Burns, S. Shnider, R. Wells, On deformations of strictly pseudoconvex domains, Inventiones Math. 46(1978), 237-253.
6. S. Chern and J. Moser, Real hypersurfaces in complex manifolds, Acta Math. 133(1974), 219-271.
7. J. D'Angelo, Real hypersurfaces, orders of contact, and applications, Ann. of Math. 115(1982), 615-637.
8. I. Graham, Boundary behavior of the Caratheodary and Kobayashi metrics on strongly pseudoconvex domains in $\mathbb{C}^n$ with smooth boundary, Trans. Am. Math. Soc. 207(1975), 219-240.

9. R. Greene and S. Krantz, Stability of the Bergman kernel and curvature properties of bounded domains, Proceedings of the Princeton Conference on Several Complex Variables, Princeton University Press, Princeton, 1981.

10. R. Greene and S. Krantz, Deformations of complex structures, estimates for the $\bar{\partial}$ equation, and stability of the Bergman kernel, Adv. Math. 43(1982), 1-86.

11. R. Greene and S. Krantz, The automorphism groups of strongly pseudo -convex domains, Math. Ann. 261(1982), 425-466.

12. R. Greene and S. Krantz, The stability of the Bergman kernel and the geometry of the Bergman metric, Bull. Am. Math. Soc. (New Series) 4(1981), 111-115.

13. R. Greene and S. Krantz, Stability of the Caratheodary and Kobayashi metrics and applications to biholomorphic mappings, Proceedings of Symposia in Pure Mathematics 41(1984), American Mathematical Society, Providence, pp. 77-93.

14. R. Greene and S. Krantz, Normal families and the semicontinuity of isometry and automorphism groups, Math. Z. 190(1985), 455-467.

15. G.M. Henkin and Leiterer, Theory of functions of strictly pseudoconvex sets with non-smooth boundary, Report of the Akadamie der Wissenschaften der DDR, Berlin, 1981.

16. L. Hörmander, The Analysis of Linear Partial Differential Operators I, Springer, Berlin, 1983.

17. N. Kerzman, Hölder and L^p solutions for the equation $\bar{\partial}u = f$ on strongly pseudoconvex domains, Comm. Pure Appl. Math. XXIV (1971), 301-380.

18. J. Kohn, Boundary behavior of $\bar{\partial}$ on weakly pseudoconvex manifolds of dimension two, J. Diff. Geom. 6(1972), 523-542.

19. J. Kohn, Global regularity for $\bar{\partial}$ on weakly pseudoconvex manifolds, Trans. A.M.S. 181(1973), 273-292.

20. J. Kohn, Subellipticity of the $\bar{\partial}$-Neumann problem on pseudo-convex domains: sufficient conditions, Acta Math. 142(1979), 79-122.

21. W. Koppelman, The Cauchy integral formula for functions of several complex variables, Bull. Am. Math. Soc. 73(1967), 373-377.

22. S. Krantz, Function Theory of Several Complex Variables, John Wiley and and Sons, New York, 1982.

23. R. Narasimhan, Several Complex Variables, University of Chicago Press, Chicago, 1971.

24. R.M. Range, The Caratheodary metric and holomorphic maps on a class of weakly pseudoconvex domains, Pac. Jour. Math. 78(1978), 173-189.

25. Reiffen, Die differnetialgeometrischen Eigenschaften der invarianten Distanzfunktion von Caratheodory, Schr. Math. Inst. Univ. 26(1963).

26. J. Rosay, Sur une characterization de la boule parmi son groupe d'automorphismes, Ann. Inst. Four. Grenbole XXIX (1979), 91-97.

27. W. Rudin, Function Theory in the Unit Ball of $\mathbb{C}^n$, Springer Verlag, Berlin, 1980.

28. T. Sunada, Holomorphic equivalence problem for bounded Reinhardt domains, Math. Ann. 235(1978), 111-128.

29. N. Tanaka, On generalized graded Lie algebras and geometric structures I, J. Math. Soc. Japan 19(1967), 215-254.

30. B. Wong, Characterization of the ball in $\mathbb{C}^n$ by its automorphism group, Invent. Math. 41(1977), 253-257.

31. H. Wu, Normal families of holomorphic mappings, Acta Math. 119(1967), 193-233.

INTERPOLATION THEORY IN $\mathbb{C}^n$: A SURVEY

Rita Saerens
Department of Mathematics
Michigan State University
East Lansing, MI 48824-1027

0. INTRODUCTION

This paper is a survey of some known results and open problems in the interpolation theory for various function algebras. First, we introduce some notations and definitions.

We will always use D to denote a bounded domain in $\mathbb{C}^n$. By $A^k(D)$ $(0 \leq k \leq \infty)$ we denote the algebra $\mathcal{O}(D) \cap C^k(\bar{D})$ of functions holomorphic on D and of class C^k up to the boundary. A closed subset K of the boundary bD is called an <u>interpolation set for</u> $A^k(D)$ if for each function f in $C^k(K)$, there exists an F in $A^k(D)$ with $F = f$ on K. It is called a <u>peak interpolation set for</u> $A^k(D)$ if we can moreover choose an interpolating function F in $A^k(D)$ with the property that $|F(x)| < \sup_K |f| = \|f\|_K$ for all x in $\bar{D} \backslash K$, whenever f is not identically equal to zero. A closed set K in bD on which the constant function 1 can be peak interpolated by a function in $A^k(D)$, is called a <u>peak set for</u> $A^k(D)$. A closely related concept we will occasionally consider is that of zero set: a compact subset K of bD is a <u>zero set for</u> $A^k(D)$ if there exists an $F \in A^k(D)$ such that $K = \{z \in \bar{D} : F(z) = 0\}$.

It is often useful to consider local versions of the above concepts. The closed subset K of bD is called a <u>local peak</u> (resp. <u>local interpolation</u>, <u>local peak interpola-</u>

tion) set for $A^k(D)$ if each point in K has an open neighborhood U in $\mathbb{C}^n$ such that $K \cap \bar{U}$ is a peak (resp. interpolation, peak interpolation) set for $A^k(D)$.

These sets have been extensively studied for the unit disc in $\mathbb{C}$. The Fatou-Rudin-Carleson theorem (see Stout [52], p. 204) states that the classes of zero, interpolation, peak and peak interpolation sets for $A(U)$ coincide and are precisely the subsets of Lebesgue measure zero of bU. For $A^k(U)$ with $k \geq 1$, the situation is quite different. For example, Taylor and Williams [54] showed that peak sets for $A^k(U)$ $(k \geq 1)$ are necessarily finite and that all finite subsets of bU are peak interpolation sets for $A^\infty(U)$. On the other hand, Carleson [8] characterized the zero sets of $A^k(U)$ $(k \geq 1)$ as the closed subsets E of bU of Lebesgue measure zero which have the property that $bU \setminus E = \bigcup I_j$ with $\sum |I_j| \log |I_j|^{-1} < \infty$, where $|I_j|$ denotes the length of the open interval I_j.

Some of the results in $\mathbb{C}^n$ are similar to those mentioned above for $\mathbb{C}$ but in general much less is known and we will mainly concentrate on the following questions:

(1) Characterize as much as possible the peak (resp. interpolation, peak interpolation) sets for $A^k(D)$.

(2) When are local peak (resp. local interpolation, local peak interpolation) sets for $A^k(D)$ global peak (resp. interpolation, peak interpolation) sets for $A^k(D)$?

(3) When are peak (resp. interpolation) sets for $A^k(D)$ necessarily peak interpolation sets or interpolation (resp. peak) sets for $A^k(D)$?

(4) When is it true that every compact subset of a peak (resp. peak interpolation) set for $A^k(D)$ is also a peak (resp. peak interpolation) set for $A^k(D)$?

The techniques used in answering the above questions for $k = 0$ or $k \neq 0$ are very different. In the case $k = 0$ we have Banach algebra techniques and results available.

For $k \neq 0$, $\bar{\partial}$-techniques play a fundamental role. This leads us to restrict our attention mostly to the polydisc U^n, strictly pseudoconvex domains and (weakly) pseudoconvex domains and to the cases $k = 0$ or ∞. Moreover, for the polydisc we consider only subsets of the distinguished boundary. (Globevnik in [27] studied more general peak sets for $A(U^n)$.)

Most of the above questions have been fairly satisfyingly answered for the polydisc and for strictly pseudoconvex domains. There are, for example, roughly speaking two types of conditions known that yield characterization results: covering type ones and geometric ones. These conditions are not equivalent but one dimensional manifolds satisfying the geometric ones often satisfy the covering type ones. The covering conditions are usually sufficient ones, while the geometric ones are often necessary and sufficient ones. Most of the above questions are far from having a satisfying answer for pseudoconvex domains.

This paper does not contain complete proofs of the results mentioned. They are all available in the literature. Sketches of some proofs are given to illustrate the techniques used.

1. THE ALGEBRA A(D)

1.1 Characterization of peak, interpolation and peak interpolation sets

For the polydisc U^n, it is well-known that for closed subsets of the distinguished boundary $\mathbb{T}^n$, the classes of interpolation, peak, peak interpolation and zero sets for $A(U^n)$ coincide (see Rudin [45], p. 132). The same result holds for strictly pseudoconvex domains. The key step consists in showing that a zero set is also a peak set. This was done by Valskii [56] for starshaped strictly pseudoconvex domains with $\mathbb{C}^3$-boundary and then for arbitrary (not necessarily simply connected) strictly pseudoconvex domains with $\mathbb{C}^2$-boundary by Chollet [15] and Weinstock [59] (see also section 1.3). Hence for the unit polydisc and strictly pseudoconvex domains we only need to study the peak interpolation sets. This is no longer true for weakly pseudoconvex domains as we will see later.

One of the most powerful tools used in the characterization of peak interpolation sets for A(D) on any domain D is Bishop's theorem (see [4]) which states that if K is a compact subset of bD on which the total variation of every annihilating measure of A(D) is zero then K is a peak interpolation set for A(D).

Davie and Øksendal [18] used Bishop's theorem to prove the following result for strictly pseudoconvex domains

Theorem : *Let* D *be a domain with* $\mathbb{C}^2$*-boundary and* K *a compact subset of* bD *such that* bD *is strictly pseudoconvex in a neighborhood of* K. *Assume that for each* $\varepsilon > 0$, K *can be covered by a sequence* $\{B(x_i,r_i)\}$ *of non-isotropic balls in* bD *with* $\sum r_i < \varepsilon$. *Then* K *is a peak interpolation set for* A(D).

(The <u>non-isotropic ball</u> $B(x,r) = \{z \in bD : |\langle z-x, \nu(x)\rangle| < r, |z-x|^2 < r\}$, where $\nu(x)$ is the outward unit normal to bD at x.)

The ideas used in proving this result are central to many other results in this area and we outline them briefly. By combining the dominated convergence theorem and Bishop's theorem the proof of this result is reduced to constructing a uniformly bounded family $\{F_\varepsilon\}_{\varepsilon>0}$ of functions in A(D) such that $\lim_{\varepsilon\to 0} F_\varepsilon(z) = 0$ for $z \in \bar{D}\setminus K$ and $|1 - F_\varepsilon(z)| < \varepsilon^{-1/2}$ for $z \in K$. For $D = \{z \in \mathbb{C}^n : \rho(z) < 0\}$, the functions F_ε are obtained as follows: for $\varepsilon > 0$, let $\{B(x_i,r_i)\}$ be the finite cover of K by non-isotropic balls with $\sum r_i < \varepsilon$ and let $F_\varepsilon(z) = 1 - \prod G_i(z)(\alpha_i r_i \varepsilon^{-1/2} + G_i(z))^{-1}$ where the $G_i \in A(D)$ are such that $G_i(x_i) = 0$, $\mathrm{Re}\, G_i(z) \geq m_i|z - x_i|^2$ on $\bar{D}$, $\mathrm{Re}\, G_i(z) \leq M_i|z - x_i|^2$ on bD , $\mathrm{grad}\,(\mathrm{Re}\, G_i)(x_i) = -\,\mathrm{grad}\, \rho(x_i)$. By choosing the α_i carefully (depending on the positive constants m_i and M_i), the functions F_ε will satisfy all of the above properties. The construction of such strong support functions G_i is rather technical for general strictly pseudoconvex domains (see [18], lemma 1 for details) ; for strictly convex domains however, one can easily check that $G_i(z) = \langle x_i - z, N(x_i)\rangle$ where $N(x_i) = \langle \partial_{\bar{z}_1}\rho(x_i), \ldots, \partial_{\bar{z}_n}\rho(x_i) \rangle$, will do.

Davie and Øksendal also showed that any compact subset of a complex tangential curve in the boundary of a strictly pseudoconvex domain has the above covering property. A manifold is called <u>complex tangential</u> if its tangent space at every point is contained in the maximal complex tangent space of bD at that point.

Rudin [46] generalized their result to higher dimensional (immersed) submanifolds. (For earlier results in that direction, see Henkin and Tumanov [31], Nagel [38], [39].)

<u>Theorem</u>: *Let* D *be a strictly pseudoconvex domain with* $\mathbb{C}^2$*-boundary ,* Ω *an open set of* $\mathbb{R}^m$ *and* $\phi : \Omega \to \mathbb{R}^m$ *a nonsingular* C^1 *-mapping with* $\langle \phi'(x)v, N(\phi(x))\rangle = 0$ *for all* $x \in \Omega$, $v \in \mathbb{R}^m$. *Then for every compact subset* K *of* Ω , $\phi(K)$ *is a peak interpolation set for* A(D) .

(Note that if $\phi(\Omega)$ is a submanifold of bD , the stated orthogonality condition is equivalent to saying that $\phi(\Omega)$ is complex tangential.)

The proof of Rudin's theorem resembles that of Davie-Øksendal's result. However, the construction of the corresponding F_ε is less straightforward even in the strictly convex case. We refer the interested reader to [46].

Nagel and Rudin [40] (see also Nagel [39] for the unit ball) studied the intersection of a peak set for A(D) for any $\mathbb{C}^1$-domain D with a transverse curve; *i.e.*, a curve in bD whose tangent space at every point is not contained in the maximal complex tangent space of bD at that point. They showed that if γ is a transverse curve of class $\mathbb{C}^2$ in bD , then any bounded holomorphic function on D has nontangential boundary values $[\mu_\gamma]$-almost everywhere, where μ_γ is the measure on bD defined by $\int f\,d\mu = \int f(\varphi(t))\,dt$ for all $f \in C(bD)$ and any parametrization φ of γ. However, it is easy to check that if $G \in A(D)$ is a peak function for some set K, then $F = \exp[i\log(1-G)]$ is a bounded holomorphic function on D with no limit along any curve in D that ends at a point of K. Hence, a peak interpolation set of a strictly pseudoconvex domain intersects any transverse curve γ in a set of $[\mu_\gamma]$-measure zero. Combinig this with Rudin's result, we have a complete characterization of those peak interpolation sets for $A(D)$ which are contained in a submanifold of the boundary of a strictly pseudoconvex domain.

Stout [53] proved that the topological dimension of any peak interpolation set for $A(D)$ for a strictly pseudoconvex domain in $\mathbb{C}^n$ is at most $n-1$ and one might ask if a converse of Rudin's result holds, namely is every peak interpolation set for $A(D)$ (locally) contained in $(n-1)$-dimensional complex tangential submanifolds of bD ? The answer is no: Henriksen [32] showed that the boundary of any smooth domain D in $\mathbb{C}^n$ contains a peak set for $A(D)$ of Hausdorff dimension $2n-1$. (Tumanov [55] constructed one of Hausdorff dimension 2.5 in the unit sphere of $\mathbb{C}^3$.) This example indicates that the question of completely characterizing all peak interpolation sets for $A(D)$ for a strictly pseudoconvex domain is far from trivial.

The covering condition for the polydisc U^n similar to Davie-Øksendal's one for strictly pseudoconvex domains is the null S-width condition due to Forelli [20]. A set E has <u>null</u> S-<u>width</u> (where S is a set of unit vectors in $\mathbb{R}^n$) if for all $\varepsilon > 0$, there exists a family $\{u_j\}$ in S and a family $\{I_j\}$ of open intervals in $\mathbb{R}$ such that $\sum |I_j| < \varepsilon$ and $E \subset \bigcup \{x \in \mathbb{R}^n : \langle x,u_j\rangle \in I_j\}$. (Recall $|I_j|$ denotes the length of I_j.)

<u>Theorem</u>: *Let* G *be a countable union of sets of null* S*-width where* S *varies over compact subsets of unit vectors in* $\mathbb{R}^n_+$. *Then, for every compact* K *of* G, $\exp(K) = \{(e^{ix_1}, \cdots, e^{ix_n}) : x \in K\}$ *is a peak interpolation set for* $A(U^n)$.

By Bishop's theorem, it is sufficient to show that $|\nu|(\exp(K)) = 0$ for all compact subsets K of G and all measures ν annihilating $A(U^n)$. It is easier to lift the problem to $\mathbb{R}^n$ that to work on $\mathbb{T}^n$. The Riesz representation theorem yields for any complex measure ν on $\mathbb{T}^n$ a measure λ on $\mathbb{R}^n$ by requiring that for all compactly supported functions $g \in C(\mathbb{R}^n)$, $\int g\, d\lambda = \int \sum g(x)[\Pi 4/x_j \sin(x_j/4)]^2\, d\nu(y)$ where the sum is taken over all $x = (x_1, \ldots, x_n) \in \exp^{-1}(y)$. It is easy to check that a measure ν annihilating $A(U^n)$ lifts to a measure λ such that $\int \exp(i\langle x,y\rangle)\, d\lambda(y) = 0$ for all $x \in \mathbb{R}^n_+$ and that hence, it suffices to prove that $|\lambda|(K) = 0$ for all compact subsets K of G and all complex Borel measures λ on $\mathbb{R}^n$ with the above property. This follows from showing that for all functions g in a suitable dense subset of $L^2(|\lambda|)$ the estimate

$$\int_{I_u} |g|^2\, d\lambda \leq C(S,g)\, \|I_u\|\, \|g\|_{L^2(|\lambda|)}$$

holds for some constant C depending on S and g and for $I_u = \{x \in \mathbb{R}^n : \langle x,u\rangle \in I\}$ and $u \in S$ (see Rudin [45], pp. 140-147 or Forelli [18]).

Geometric results similar to those of Rudin also exist (see Saerens [48]). The submanifolds of the distinguished boundary $\mathbb{T}^n$ that occur are those which satisfy one of the cone conditions introduced by Burns and Stout [7]: a submanifold M of $\mathbb{T}^n$ satisfies the <u>closed cone condition</u> if at every point of M the tangent space to M does not intersect the open positive cone in the tangent space of $\mathbb{T}^n$ at that point. It satisfies the <u>open cone condition</u> if at every point of M the tangent space of M intersects the closed positive cone in the tangent space of $\mathbb{T}^n$ trivially. As was the case for the geometric and covering type conditions for strictly pseudoconvex domains, curves in $\mathbb{T}^n$ satisfying the open cone condition also satisfy the null S-width condition of Forelli's result. This need not be true for higher dimensional submanifolds (see Saerens [48]). The analogue to Rudin's result is as follows:

<u>Theorem</u> : *Let* M *be a* C^2*-submanifold of* $\mathbb{T}^n$ *which satisfies the open cone condition. Then every compact subset of* M *is a peak interpolation set for* $A(U^n)$*. Conversely, if* M *is such that every compact subset of it is a peak interpolation set for* $A(U^n)$ *then* M *satisfies the closed cone condition.*

(As was the case with Rudin's result, this theorem holds for immersed manifolds.)

The key remark in the proof of the first part is to note that the manifold M can be mapped, locally at any point p of M, into a complex tangential manifold of class C^1 in the boundary of the unit ball of $\mathbb{C}^n$. Using this mapping, Rudin's proof for the similar result on the ball can be transcribed to the polydisk. The proof of the second part is similar to the Nagel-Rudin result mentioned before: A manifold M which does not satisfy the closed cone condition has the property that for every point p of M, there is a nontangential curve $\gamma_p : [0,1) \to U^n$ (*i.e.*, a curve whose projections in each coordinate plane are nontangential) with $\lim_{t\to 1} \gamma_p(t) = p$ such that for every bounded holomorphic function F on U^n, $\lim_{t\to 1} F(\gamma_p(t))$ exists for almost all p in M (with respect to the natural measure on M). Hence by the remark made previously, compact subsets of such an M cannot be peak interpolation sets for $A(D)$.

Problem : Find necessary and sufficient conditions for a (smooth) submanifold of $\mathbb{T}$ to have the property that all its compact subsets are peak interpolation sets for $A(U^n)$.

For pseudoconvex domains much less is known about the characterization of peak, interpolation and peak interpolation sets for $A(D)$. It is not true (as it was for strictly pseudoconvex domains and for the polydisc) that the classes of these sets coincide (see also section 3). Since for any domain D with C^2-boundary the peak points for $A(D)$ are contained in the closure of the set of strictly pseudoconvex boundary points (see Basener [2]), not every interpolation set for $A(D)$ is a peak set for $A(D)$ even if the domain is pseudoconvex. The question of when every boundary point of a pseudoconvex domain is a peak point for $A(D)$ is far from completely answered. Bedford and Fornaess [3] proved the following

Theorem : *Every boundary point of a domain* D *in* $\mathbb{C}^2$ *is a peak point for* $A(D)$ *if* D *is*

(a) *a pseudoconvex domain with real analytic boundary or*

(b) *a pseudoconvex domain of finite type with* C^∞*-boundary.*

If D *is a pseudoconvex domain with real analytic boundary in* $\mathbb{C}^n$ *then boundary points where the Levi form has corank* 1 *are peak points.*

A domain D in $\mathbb{C}^2$ is of <u>finite type</u> if there is a finite bound on the order of contact one-dimensional complex manifolds have with bD. In $\mathbb{C}^n$, there are several distinct concepts of "finite type" (see for example Bloom [5] and d'Angelo [17]).

Problem : Let D be a smoothly bounded pseudoconvex domain in $\mathbb{C}^n$ for which the supremum of the order of contact one-dimensional complex varieties (or complex hypersurfaces) have with bD is finite. Is then every boundary point of D a peak point for A(D)?

Bedford and Fornaess [3] proved moreover that in the case of a pseudoconvex domain in $\mathbb{C}^2$ with real analytic boundary, the peak functions can be chosen to vary continuously with the peak point. This was generalized by Fornaess and Krantz [23] to any compact metric space X and closed subalgebra of $C(X)$.

Because of the above remarks it is clear that complex tangential manifolds cannot play the same role in the interpolation theory for A(D) of pseudoconvex domains as they did for strictly pseudoconvex domains. However, using Whitney covers, del Castillo [9] proved a result very similar to the covering result of Davie and Øksendal [18] for smoothly bounded convex domains containing no line segment in their boundaries. A <u>Whitney cover</u> for a subset E of bD is a family $\{B(x_j,r_j)\}$ of non-isotropic balls which are pairwise disjoint and for which there exists constants k and h such that E is contained in the union of the $B(x_j,kr_j)$, every point of E belongs to at most M differrent balls $B(x_j,kr_j)$ and every $B(x_j,hr_j)$ intersects bD\E. (Any open set in the boundary of a domain D with C^2-boundary has a Whitney cover, see Coifman and Weiss [16], p.70.)

Theorem : *Let* D *be a convex domain with a* C^2*-boundary not containing line segments. Let* K *be a closed subset of* bD *with the properties that* K *has zero area measure and* bD\K *has a Whitney cover* $\{B(x_i,r_i)\}$ *with* $\sum r_j < \infty$. *Then* K *is a zero set and a peak set for* A(D).

The proof bears a strong resemblance to that of a theorem of Chollet in [14] which states that a similar metric condition on a Whitney cover for $bD\setminus K$ when D is strictly pseudoconvex, implies that K is a zero set for $A^\infty(D)$ (see section 2.1). The properties on D and the fact that $\mathrm{Re}\,\langle x_i - z, \nu(x_i)\rangle > 0$ on $\bar{D}\setminus\{x_i\}$ imply that the function

$$\phi(z) = \sum \lambda_i r_i \left(r_i + \langle x_i - z, \nu(x_i)\rangle\right)^{-1}$$

where λ_i are positive numbers tending to infinity and such that $\sum \lambda_i r_i < \infty$, is holomorphic on D, continuous on $\bar{D}\setminus K$ and tends to infinity whenever $z \in bD\setminus K$ tends to a point of K. The function $\exp(-\phi(z))$ extends to a function $F \in A(D)$ which vanishes exactly on K. Since convex domains are simply connected, zero sets for $A(D)$ are clearly also peak sets.

Both Davie-Øksendal's and del Castillo's constructions work on any domain D for which there exists a function G defined on $bD \times \bar{D}$ such that $G(x,\cdot) \in A(D)$ for all x in bD, $G(x,x) = 0$, $\mathrm{grad}\,(\mathrm{Re}\,G(x,x)) = -\,\mathrm{grad}\,\rho(x)$, $\mathrm{Re}\,G(x,z)$ is estimated from below on $b\bar{D} \times \bar{D}$ by $m|\zeta - z|^2$ or $m|\langle z-x, \nu_x\rangle|$ for some positive m and from above on $bD \times bD$ by $M|\zeta - z|^2$ or $M|\langle z-x, \nu_x\rangle|$ for some positive M.

Problem : For which domains does such a function H exist ?

del Castillo also shows that any compact subset of a complex tangential curve of class C^2 satisfies the above conditions, which leads to

Problem : Is it true that every closed subset of a complex tangential submanifold of bD is a peak set for $A(D)$ when D is a smoothly bounded convex domain not containing line segments in its boundary?

del Castillo's result illustrates the importance of a suitable generalization of the concept "non-isotropic ball" to more general pseudoconvex domains.

1.2. Local versus global

Since finite unions of peak sets for A(D) are clearly peak sets, local peak sets are always global peak sets.

Problem : Let D be a pseudoconvex domain, are local interpolation sets for A(D) necessarily global ones ? (Same question for local peak interpolation sets.)

1.3. Implications

It is clear from the definitions that any peak interpolation set for A(D) is also an interpolation, peak and zero set. What other such implications hold for pseudoconvex domains (recall the remarks made in 1.1) ?

We have the following result due to Varopoulos [57] which emphasizes the importance of studying when every boundary point is a peak point for A(D).

Theorem : *Let* D *be any bounded domain in* $\mathbb{C}^n$ *and* K *a compact interpolation set for* A(D) *of which every point is a peak point for* A(D). *Then* K *is a peak interpolation set for* A(D).

Glicksberg [26] gave an easy proof of this theorem along the following lines: since K is an interpolation set, the open mapping theorem guarantees that there exists a positive constant α such that every $f \in \mathcal{C}(K)$ extends to an $F \in A(D)$ with $\|F\|_D \leq \alpha\|f\|_K$. Fix now $\varepsilon > 0$, a measure ν which annihilates A(D) and an open set Ω with $K \subset \Omega$ and $|\nu|(\Omega \setminus K) < \varepsilon$. Since every point of K is a peak point for A(D), one can choose functions $f_1, \ldots, f_N$ in A(D) such that $\|f_j\| = 1$, $|f_j(z)| < \varepsilon$ for $z \in \overline{D} \setminus \Omega$ and that the sets $E_j = \{z \in D : |1-f_j(z)| < \varepsilon\}$ cover K. Let K_j be pairwise disjoint compact sets with $K_j \subset E_j \cap K$ and $|\nu|(\Omega \setminus \cup K_j) < 2\varepsilon$. For each j, choose functions $g_k \in A(D)$ with $\|g_k\| \le \alpha$ and $g_k(z) = \omega^{kj}$ on K_j, where ω is a fixed N^{th}-root of unity. Since the functions $h_m = N^{-1} \Sigma \omega^{-km} g_k$ are one on K_m and zero on K_j with $j \neq m$, the function $H = \Sigma h_m{}^2 f_m$ is such that $\|H\| \le \alpha^2$, $|1 - H(z)| < \varepsilon$ on $\cup K_j$ and $|H(z)| \le \alpha^2\varepsilon$ on $\bar{D} \setminus \Omega$. Repeating this construction for $\varepsilon_n \to 0$ yields a uniformly bounded sequence of functions $\{H_n\}$ in $A(D)$ which converges pointwise to the characteristic functions of K. The conclusion of the theorem follows then from Bishop's theorem.

Jimbo [36] studied the question as to which conditions on the set of weakly pseudoconvex domain would imply that a peak set for $A(D)$ is necessarily a peak interpolation set. The proofs he gives make use of a theorem of Fornaess and Nagel [24] which is not correct under the generality stated. Jimbo's proof is correct whenever the theorem of Fornaess and Nagel holds (for example, when the set of weakly pseudoconvex boundary points is totally real or when D is in $\mathbb{C}^2$ with a real analytic boundary). Since no counterexample to Jimbo's result is known, we state it as a problem.

__Problem__ : PROVE OR DISPROVE : *Let* D *be a pseudoconvex domain with* C^∞*- boundary in* $\mathbb{C}^n$. *Assume that the set of weakly pseudoconvex boundary points consists of peak points for* $A(D)$ *and is a stratified totally real set. Then every peak set for* $A(D)$ *is a peak interpolation set for* $A(D)$.

A set is called a __stratified totally real set__ if it is the disjoint union of sets M_j where each M_j is a totally real C^∞-manifold of $\mathbb{C}^n \setminus \cup_{i<j} M_i$.

__Problem__ : If the above statement is false, find other conditions on the domain for which such a result holds.

For simply connected domains a zero set for $A(D)$ is always a peak set since if f is a function with $\|f\| < 1$, $\exp(1 - \log f(1-\log f)^{-1})$ is equal to one on $\{z : f(z) = 0\}$ and strictly less than one elsewhere. Chollet [15] obtained the same result for an arbitrary strictly pseudoconvex domain D with C^2-boundary by constructing at every point $p \in bD$, a simply connected strictly pseudoconvex domain $D_p \subset D$ such that bD_p and bD agree near p. This yields a local peak function (on D_p) which can be extended to D by standard $\bar{\partial}$-techniques. These $\bar{\partial}$-techniques do not hold on general pseudoconvex domains. Verdera [58] obtained the same result for pseudoconvex domains by using an additive Cousin problem.

1.4. Compact subsets of peak sets and peak interpolation sets

Compact subsets of interpolation (resp. peak interpolation) sets are clearly again interpolation (resp. peak interpolation) sets. By the remark made in 1.1 the same holds for peak sets in the case of the unit polydisc or a strictly pseudoconvex domain. It seems to be still an open question whether this is true for pseudoconvex domains.

2. THE ALGEBRA $A^\infty(D)$

2.1 Characterization of peak, interpolation and peak interpolation sets

Compact subsets of complex tangential submanifolds of the boundary of the unit ball B are in general not peak interpolation sets for $A^\infty(B)$ (see Saerens [48]). Using Hopf's lemma, Noell [43] showed

Theorem : *Let* D *be a bounded domain with* C^2 *-boundary and* K *a peak interpolation set for* $A^k(D)$ $(1 \le k \le \infty)$. *Then* K *is finite.*

Problem : Characterize which finite subsets are peak interpolation sets for $A^k(D)$ when D is a (strictly) pseudoconvex domain.

There are no results known for differentiable peak interpolation on the polydisc which yields immediately the following questions:

Problem : Are closed subsets of manifolds in $\mathbb{T}^n$ which satisfy the open cone condition (local) peak interpolation sets for $A^k(U^n)$? Are there any infinite peak interpolation sets for $A^k(U^n)$?

For strictly pseudoconvex domains, complex tangential submanifolds of the boundary play an important role in the study of peak and interpolation sets for $A^\infty(D)$ just like they did for $A(D)$. We have the following result:

Theorem : *Let* D *be a strictly pseudoconvex domain with* C^∞ *-boundary and* M *a complex tangential submanifold of class* C^∞ *of* bD. *Then every closed subset* K *of* M *is a peak set and an interpolation set for* $A^\infty(D)$.

Hakim and Sibony [30] proved the interpolation and local peak results. They follow from the following technical lemma: For every $p \in K$, there exist neighborhoods U and V of p, with V compactly contained in U, and a function $\varphi \in C^\infty(U)$ with the following properties : (i) $\varphi = 0$ on $K \cap V$; (ii) $\mathrm{Re}\, \varphi < 0$ on $(\bar{D} \cap U) \setminus (K \cap V)$; (iii) $D^\alpha(\bar\partial\varphi) = 0$ on $K \cap V$ for all α ; (iv) $\bar\partial(1/\varphi)$ extended to $K \cap V$ by zero belongs to $C^\infty(\bar{D} \cap U)$; (v) for every $F \in C^\infty(\bar{D} \cap U)$ with $D^\alpha(\bar\partial F) = 0$ for all α, $(1/\varphi)\bar\partial F$ extended to $K \cap V$ by zero, belongs to $C^\infty(\bar{D} \cap U)$. (It is in the proof of this lemma that the condition on the tangent space of M is used.) The local peak result follows easily. Indeed, choose a C^∞-function χ with compact support in U and equal to one on $\bar{V}$. Let g be a solution of $\bar\partial u = \bar\partial(\chi/\varphi)$. Then $h = \varphi / (\chi - g\varphi)$ is zero on $K \cap V$ and $\mathrm{Re}\, h < 0$ on

$\bar{D} \setminus (K \cap V)$. The local interpolation result is obtained similarly. The global one follows from this by a partition of unity argument. The global peak result does not follow so readily from the local one since finite unions of peak sets for $A^{\infty}(D)$ are not always peak sets (see Hakim and Sibony [30]). Chaumat and Chollet [11] obtained the global peak result by showing that for any compact subset K of a complex tangential manifold M there exist a neighborhood U of K in $\mathbb{C}^n$, a strong support function $\varphi \in C^{\infty}(U)$ and a positive constant c such that (i) $K = \{z \in U : \varphi(z) = 0\}$; (ii) $D^{\alpha}(\bar{\partial}\varphi) = 0$ on $U \cap M$ for all α; (iii) $\mathrm{Re}\, \varphi(z) \geq c\,[\mathrm{dist}(z,M)]^2$ for all $z \in U \cap \bar{D}$. To obtain a peaking function on K, pick a C^{∞}-function ψ with support in U such that $0 \leq \psi \leq 1$ and $\psi = 1$ in a neighborhood V of K. Then $\bar{\partial}(\psi/\varphi)$ extends to a closed (0,1)-form of class C^{∞} in D. Let u be a C^{∞}-solution of $\bar{\partial}u = \bar{\partial}(\psi/\varphi)$. By adding a positive constant to the holomorphic function $v = \psi/\varphi - u$, we may assume $\mathrm{Re}\, v > 0$ on $\bar{D} \setminus K$ and hence $\exp(-1/v)$ peaks on K.

Chaumat and Chollet [12] also obtained a partial converse

Theorem : *Let* D *be a strictly pseudoconvex domain with* C^{∞}*-boundary and let* K *be a closed subset of* bD *which is a local peak set for* $A^2(D)$. *Then* K *is locally contained in a complex tangential submanifold of* bD.

The above result raises the question whether global peak sets are globally contained in complex tangential manifolds. Chaumat and Chollet [13] showed the answer is negative for $n \geq 4$. They construct a compact K of some open set U of $\mathbb{C}^3$ such that K is the zero set of a positive plurisubharmonic function in $C^{\infty}(U)$ and is not contained in any proper submanifold of U. The compact K is then naturally imbedded in the boundary of a strictly pseudoconvex domain D in $\mathbb{C}^4$ such that K is locally but not globally contained in complex tangential submanifolds of bD. Fornaess and Henriksen [22] proved that for $n = 3$, peak sets for $A^{\infty}(D)$ are always globally contained in complex tangential submanifolds by first proving the existence of a stratification by complex tangential manifolds whose union contains K. When $n = 3$, this stratification yields the existence of a complex tangential manifold containing K.

The polydisc case shows the same simililarity between $k = \infty$ and $k = 0$ as the strictly pseudoconvex one does (see Saerens [47]).

Theorem : *Let* M *be a* C^∞*-submanifold of the distinguished boundary of the unit polydisc* U^n *which satisfies the open cone condition. Then every closed subset of* M *is a local peak set and an interpolation set for* $A^\infty(U^n)$.

The local interpolation and peak results are obtained with the same methods as in the case of $A(U^n)$: M is locally mapped into a smooth complex tangential manifold in the boundary of some strictly pseudoconvex domain with C^∞-boundary. The global interpolation result is obtained by a partition of unity argument. This embedding method does not however yield a global peak result. A complete characterization of peak sets for $A^\infty(U^n)$ is known.

Theorem : *Let* K *be a compact subset of the distinguished boundary* $\mathbb{T}^n$ *of the unit polydisc* U^n. *Then the following are equivalent*

(a) K *is a peak set for* $A^\infty(U^n)$;

(b) *there is an open neighborhood* Ω *of* K *in* $\mathbb{T}^n$ *and a closed* C^∞*- submanifold* M *of* Ω *which satisfies the open cone condition and contains* K.

The fact that (a) implies (b) was proved by Saerens and Stout [49] by using Hopf's lemma. It was also shown there that it might not be possible to take for M a closed submanifold of the whole $\mathbb{T}^n$. As mentioned earlier a local version of the converse implication was proved by Saerens [48]. Labonde [37] proved the global peak result by using $\bar{\partial}$-techniques on the polydisc similar to those used by Chaumat-Chollet [11] in the case of strictly pseudoconvex domains.

We recall (see section 1.1) that Bedford and Fornaess [3] showed that every boundary point of a pseudoconvex domain D in $\mathbb{C}^2$ with either a C^∞-boundary and of finite type or with real analytic boundary is a peak point for $A(D)$. This is in some sense the best possible result for such domains: Fornaess [21] constructed a pseudoconvex domain with real analytic boundary which has only one weakly pseudoconvex point and such that it is not a peak point for $A^1(D)$. The study of the existence of differentiable peak functions requires looking at stronger "finite type" conditions than those defined in section 1.1 : Let 0 be a boundary point of a smoothly bounded domain D in $\mathbb{C}^n$ with defining function ρ and assume the tangent space to bD at 0 is given by $\operatorname{Re} z_n = 0$. Then 0 is a point of <u>strict finite type</u> if there are a function h holomorphic in a neighborhood V of 0 in $\mathbb{C}^{n-1}$, a positive integer k and a positive constant m such that $h(0) = 0$ and $\rho(w,h(w)) \geq m|w|^k$ on V. Hakim and Sibony [29] showed that a boundary point p of strict finite type is a local peak point for $A^1(D)$ (*i.e.*, there is a neighborhood U of p in such that some function F in $A^1(D \cap \bar{U})$ peaks at p). Bloom [5] showed that this strict finite type condition is not sufficient to yield an $A^\infty(D)$ peaking result. (He also introduced a stronger type condition necessary and sufficient for a boundary point of a pseudoconvex domain to be a local peak point for functions holomorphic in a neighborhood of the point.)

Problem : Are there type conditions that are necessary and sufficient for a point to be a (local) peak point for $A^\infty(D)$?

Noell [41] showed that for $t > 0$ sufficiently small, the convex domain $D = \{(z,w) \in \mathbb{C}^2 : \operatorname{Re} w + |z|^4 - t(\operatorname{Im} z)^4 + (\operatorname{Im} w)^2 < 0\}$ near 0 has the property that $M = \{(z,w) \in bD : \operatorname{Im} z = \operatorname{Im} w = 0\}$ is a complex tangential curve such that $M \cap \bar{U}$ is not a peak set for $A^\infty(D)$ for any neighborhood U of 0. He also pointed out that for $D = \{(z,w) \in \mathbb{C}^2 : |z|^4 + |w|^2 < 1\}$ the set $K_1 \cup K_2$ where $K_1 = \{(z,w) \in bD : \operatorname{Im} z = \operatorname{Im} w = 0\}$ and $K_2 = \{(z,w) \in bD : \operatorname{Re} z = \operatorname{Im} w = 0\}$ is a peak set for $A^\infty(D)$ which is not (locally) contained in any complex tangential submanifold of bD. Hence complex tangential manifolds do not play here the same role as they do for strictly pseudoconvex domains. Complex tangential conditions at certain points however still do play a role. Iordan [33] obtained some necessary conditions for local peak sets for $A^\infty(D)$ involving them.

Theorem : *Let* D *be a domain with* $\mathcal{C}^2$*-boundary,* K *a local peak set for* $A^2(D)$ *and* p *a point in* K *where* bD *is pseudoconvex and where the Levi form has* q *zero eigenvalues. Then there is a neighborhood* V *of* p *and an* $(n+q-1)$*-dimensional* $\mathcal{C}^1$*-submanifold* M *of* $V \cap bD$ *such that*

(a) $K \cap V$ *is contained in* M;

(b) M *is complex tangential at all points* $x \in K$;

(c) *the complex dimension of the maximal complex tangent space of* M *at any point is at most* q.

The first example of Noell mentioned above also shows that the necessary conditions in Iordan's theorem are not sufficient. Iordan proved an analogue of Hakim and Sibony's technical lemma mentioned before in this section and used it obtain sufficient conditions for a set to be a local peak set for $A^{\infty}(D)$ in terms of strictly q-convexity and the existence of CR-functions with certain growth estimates (see [33] for details).

2.2 Local versus global

Fornaess and Henriksen [22] showed that for a strictly pseudoconvex domain D with $\mathcal{C}^{\infty}$-boundary local peak sets for $A^{\infty}(D)$ are always global peak sets by proving the following technical lemma: Let M_1, M_2 be complex tangential submanifolds of bD with $\dim M_1 < \dim M_2$ and $M_1 \cap M_2$ open in M_1. Then any compact subset K of $M_1 \cup M_2$ such that $K \cap M_1$ is open in K, is a peak set for $A^{\infty}(D)$. The peak function for K is obtained by "gluing" the peak functions for $K \cap M_1$ and $K \cap M_2$. Using this lemma repeatedly yields that local peak sets are global ones.

This argument cannot be generalized to arbitrary pseudoconvex domain. This was shown by Noell [41]: the domain $D = \{(z,w) \in \mathbb{C}^2 : \rho(z,w) = |w+e^{i\ln z\bar{z}}|^2 - 1 + C(\ln z\bar{z})^4\}$ (where C is a large positive constant) is a pseudoconvex domain such that its set of weakly pseudoconvex boundary points $\{(z,w) \in bD : |z| = 1, w = 0\}$ is a local peak set for $A^{\infty}(D)$ but not a global one. The reason the set fails to be a peak set is because there are real analytic complex tangential manifolds contained in the set of weakly pseudoconvex boundary points. A pseudoconvex domain D in $\mathbb{C}^n$ with C^∞-boundary has the (NP)-property if the set of weakly pseudoconvex points of bD is contained in finitely many real analytic curves and does not contain any real analytic, complex tangential manifolds. For example, any convex domain with real analytic boundary in $\mathbb{C}^n$ has the (NP)-property if $n = 2$ (see Noell [41]); this is false for $n \geq 3$.

Theorem : *Let* D *be a pseudoconvex domain with real analytic boundary in* $\mathbb{C}^2$ *which has the* (NP) *-property. Then any local peak set for* $A^\infty(D)$ *is a peak set for* $A^\infty(D)$.

Noell [41] proved this by patching local peak funcions. This patching cannot always be done at weakly pseudoconvex points but the hypotheses on the domain guarantee it need only be done at strict pseudoconvex points.

Problem : Find conditions for $n \geq 3$ which imply the same result.

The question of whether local peak sets for $A^\infty(U^n)$ are necessarily global ones, is still open.

As mentioned in section 2.1, Hakim and Sibony [30] used a partition of unity argument to obtain global interpolating functions from local ones in the case of smoothly bounded strictly pseudoconvex domains (see Saerens [48] for the polydisc case). This argument works for any domain D and compact subset K of bD which is a local interpolation set and has the property that every compact subset of K is a peak set for $A^\infty(D)$.

2.3 Implications

For the function algebras $A^\infty(D)$ of a smoothly bounded domain, peak sets or interpolation sets are rarely peak interpolation sets since as we saw in section 1 the peak interpolation sets for $A^k(D)$ (for $1 \le k \le \infty$) are necessarily finite sets (see Noell [43]), while peak or interpolation sets need not be finite.

Problem : Is there any general result of the following form possible for the polydisc: If K is a peak set and an interpolation set for $A^\infty(U^n)$, then it is a peak interpolation set for $A^\infty(U^n)$?

There remain the questions whether being a peak set for $A^\infty(D)$ implies being an interpolation set for it and conversely, whether being an interpolation set for $A^\infty(D)$ implies being a peak set. For strictly pseudoconvex domains, the answer is affirmative to both questions because of the characterization of these sets given before. For the unit polydisc, it is known that if the subset K of the distinguished boundary is a local peak set for $A^\infty(U^n)$ then it is an interpolation set for $A^\infty(U^n)$ (see Saerens and Stout [49]). If the domain D is pseudoconvex only partial results are known. Interpolation sets for $A^\infty(D)$ need not be peak sets even for $A(D)$ since not all boundary points are peak points. Some conditions are known under which being a peak set implies being an interpolation set. They are due to Noell [42].

Theorem : *Let* D *be a pseudoconvex domain in* $\mathbb{C}^2$ *with* C^∞*-boundary and of finite type and let* M *be a* C^∞*-curve in the boundary which is a local peak set for* $A^\infty(D)$.

(a) *If* bD *is of constant type along* M, *then every closed subset of* M *is an interpolation set for* $A^\infty(D)$.

(b) *If* bD *and* M *are real analytic, then every closed subset of* M *is an interpolation set for* $A^\infty(D)$.

The proof of this theorem is similar to the proof of the interpolation result of Hakim and Sibony for strictly pseudoconvex domains discussed in section 2.1. This sort of argument requires the existence of a strong support function φ for which $\mathrm{Re}\varphi$ vanishes only to finite order along M. Such functions always exist when the type of bD along M is constant. If the order is not constant this need not be the case. Indeed, for $D = \{(z,w) \in \mathbb{C}^2 : |z|^4 + |w|^2 < 1\}$ the set $M = \{(z,w) \in bD : \mathrm{Im}z = \mathrm{Im}w = 0\}$ is a peak set for $A^\infty(D)$ but any strong support function vanishes to infinite order at the points $(0,\pm 1)$ in the direction of $\{(z,w) \in bD : \mathrm{Re}z = \mathrm{Im}w = 0\}$. These points $(0,\pm 1)$ are of type 4 while all the other points of M are of type 2. The above argument can be modified for points near which the type is not constant under the additional hypothesis that both bD and M are real analytic.

Noell also gives an example of a convex domain with C^∞-boundary whose set of weakly pseudoconvex boundary points is a line segment which is a peak set for $A^\infty(D)$ but is not a local interpolation set for $A^\infty(D)$. This set consists of points of infinite type.

<u>**Problem**</u> : Find similar results for $\mathbb{C}^n$ with $n \geq 3$.

Recall (see section 2.1) that for strictly pseudoconvex domains (resp. the unit polydisc) a compact subset of the boundary (resp. the distinguished boundary) is a local peak set for $A^\infty(D)$ if and only if it is locally contained in a complex tangential submanifold (resp. a submanifold satisfying the open cone condition). This is not true for pseudoconvex domain and Noell's result indicates that being a local peak set is the key factor in obtaining interpolation results (see also Saerens and Stout [49] for the polydisc case).

2.4 Compact subsets of peak sets and interpolation sets

Compact subsets of interpolation sets are clearly again interpolation sets.

The question whether closed subsets of peak sets are again peak sets, is in general less trivial. By the characterizations given of such sets in section 2.1 the answer is clearly affirmative for strictly pseudocovex domains and for the unit polydisc. For pseudoconvex domains, this is not always the case. Noell [44] constructed an example of a convex domain D with $\mathbb{C}^\infty$-boundary for which the set K of weakly pseudoconvex boundary points is a peak set for $A^\infty(D)$ but it contains a point that is not a peak point for any $A^k(D)$, $(k>0)$. His example is such that the set K is actually a line segment. Positive results due to Noell and to Iordan are known under various additional conditions on the domain to guarantee the above phenomenon cannot happen.

Theorem: *Let* D *be a pseudoconvex domain. Then* D *has the property that any closed subset* L *of a peak set* K *for* $A^\infty(D)$ *is also a peak set for* $A^\infty(D)$ *if* D *satisfies any of the following properties:*

(a) D *is contained in* $\mathbb{C}^2$ *and has a real analytic boundary* (see Noell [41]);

(b) D *is smoothly bounded and of finite type in* $\mathbb{C}^2$ (see Noell [41]);

(c) D *has a* $\mathbb{C}^\infty$ *-boundary and the* (NP) *-property* (see Iordan [35]).

For the proof of this theorem one first notices that the methods used by Chaumat and Chollet in [12] for strictly pseudoconvex domains can be used here to show that the set $L_1 = (K \cap w(bD)) \cup L$ is a peak set, where $w(bD)$ is the set of weakly pseudoconvex points of D. To prove (c) Iordan shows first that if p is an isolated point in $(K \cap w(bD)) \setminus L$, $L_1 \setminus \{p\}$ is also a peak set. This can also be used to eliminate points of $K \cap w(bD)$ near which $w(bD)$ is contained in real analytic curves which are not complex tangential at p. Indeed, such points have a neighborhood U such that $K \cap w(bD) \cap U$ is finite. Hence only non-isolated points p of $K \cap w(bD) \cap \gamma$ for a real analytic curve γ

complex tangential at p remain. The (NP)-property ensures that p is the limit (from both sides) of a sequence of points of γ at which γ is not complex tangential. Deleting appropriate neighborhoods of such points using the previous step yields a peak set L_2 such that p is an isolated point of $L_2 \cap \gamma$. If p belongs to another such curve, the procedure is repeated; otherwise p can be excised from L_2 by the previous step. Noell proved (a) by combining similar "cutting out" ideas with the fact that w(bD) can be decomposed in a union of pairwise disjoint real analytic manifolds S_0, S_1 and S_2 with the properties that (i) each S_j consists of finitely many j-dimensional real analytic totally real manifolds; (ii) S_1 is closed in $bD \setminus S_0$ and S_2 is closed in $bD \setminus (S_0 \cup S_1)$; (iii) each component of S_2 consists of points of the same finite type. This decomposition result due to Fornaess and Øvrelid [25] is only available for domains with real analytic boundaries. Noell used another decomposition theorem due to Catlin [10] for domains of finite type.

Problem : Find other sufficient conditions on D for the above property to hold (especially for domains in $\mathbb{C}^n$ with $n \geq 3$).

3. MISCELLANEOUS

For the sake of completeness we mention without any details some other interesting interpolation questions and results.

3.1 The algebras $A^k(D)$ with $0 < k < \infty$

When interpolating C^k-functions one has to expect to lose some degree of smoothness (see Rudin [47] and Stein [50]). Therefore we introduce the following concept: a closed subset K of the boundary of a domain D is an (s,k)-interpolation (resp. (s,k)-peak interpolation) set if every C^s-function (different from zero) on K can be interpolated (resp. peak interpolated) by functions in $A^k(D)$.

Noell's result state that for any bounded domain with C^2-boundary, the (k,k)-peak interpolation sets (for $1 \leq k \leq \infty$) are necessarily finite sets (see Noell [43]).

Problem : Characterize the (s,k)-peak interpolation sets.

For the polydisc we have (see Saerens [48] and Saerens and Stout [49]) that closed subsets of a C^∞-submanifold of $\mathbb{T}^n$ satisfying the open cone condition are $(s,s-1)$-interpolation sets.

The methods used by Hakim and Sibony [30] for $k = \infty$ yield that compact subsets of smooth complex tangential manifolds are (s,k)-interpolation sets, where $k = [s/2] - 1$ if s is odd and $k = [s/2] - 2$ if s is even. It is known that a loss of the order of half the derivatives is unavoidable in this situation (see Rudin [47] and Saerens [48]). In the case of a pseudoconvex domain the loss of differentiability is even bigger.

Problem : Describe the minimal loss of differentiability in terms of the geometry of the domain.

The example that Noell [41] gave to show that local peak sets for $A^\infty(D)$ need not be global peak sets can be strengthen to show that local peak sets for $A^\infty(D)$ need not even be peak sets for $A^2(D)$.

Problem : Are local peak sets for $A^k(D)$ global peak sets if the domain has the (NP)-property ?

Are there weaker conditions on the domain which yield this ?

Related to this is the interpolation of functions which are C^k and which derivatives of order k satisfy Lipschitz conditions of order m. This was studied by Saerens and Stout in [49] for the polydisc and by Bruna and Ortega in [6] for the ball.

3.2 Norm preserving interpolation

Another interpolation concept which might be useful to study is norm preserving interpolation. A closed set K in the boundary of a domain D is called a norm preserving interpolation set for $A^k(D)$ if for every function f in $C^k(K)$ there exists a function F in $A^k(D)$ which equals f on K and with $|F(z)| \leq \sup_K |f|$ for all $z \in \bar{D}\setminus K$. This has been studied for $k = 0$ by Globevnik [28] in the case of the polydisc.

Problem : Study the norm preserving interpolation sets for $k > 0$ for the polydisc.

What can be said about norm preserving interpolation sets for (strictly) pseudoconvex domains ?

3.3 Maximum modulus sets

We call a closed subset K in the boundary of a domain D a maximum modulus set for $A^k(D)$ if there exists a function f in $A^k(D)$ such that $|f| = 1$ on K and $|f| < 1$ on $\bar{D} \setminus K$. (The concept of local maximum modulus set is defined similarly.) Duchamp and Stout [19] showed that if D is a strictly pseudoconvex domain in $\mathbb{C}^n$ and M is an n-dimensional C^2-submanifold of bD which is a maximum modulus set for $A^2(D)$ then M is totally real and foliated by compact complex tangential manifolds. They also have a partial converse. Iordan [34] showed that a closed subset of the boundary of a strictly pseudoconvex domain D which is a local maximum modulus set for $A^\infty(D)$ is locally contained in a totally real manifold which is foliated by complex tangential manifolds.

Problem : What can be said about (local) maximum modulus sets for pseudoconvex domains and for the polydisc ?

3.4 Interpolation of order s

A natural concept in the interpolation theory of C^∞-functions is that of interpolation of order s $(0 \leq s \leq \infty)$. A closed subset K of bD is an interpolation set of order s if for each function f in $C^\infty(K)$ with $\bar{\partial} f = 0$ vanishing to order s-1 on K there is a function F in $A^\infty(D)$ such $D^\alpha F = D^\alpha f$ on K for all multi indices α with $0 \leq |\alpha| \leq s$. Chaumat and Chollet [12] showed that for a strictly pseudoconvex domain a peak set for $A^\infty(D)$ is an interpolation set of order s for all $0 \leq s \leq \infty$. Noell's result for pseudoconvex domains mentioned in section 2.3 also gives interpolation of infinite order.

4. REFERENCES.

[1] Y. ABE, A necessary condition for the existence of peak functions, *Mem. Fac. Sci. Kyushu Univ.* Ser. A 37 (1983), 1-8.

[2] R. F. BASENER, Peak points, barriers and pseudoconvex boundary points, *Proc. Am. Math. Soc.* 65 (1977), 89-92.

[3] E. BEDFORD and J. E. FORNAESS, A construction of peak functions on weakly pseudoconvex domains, *Ann. Math.* II Ser. 107 (1978), 555-568.

[4] E. BISHOP, A general Rudin-Carleson theorem, *Proc. Am. Math. Soc.* 13 (1962), 140-143.

[5] T. BLOOM, C^∞ peak functions for pseudoconvex domains of strict type, *Duke Math. J.* 45 (1978), 133-147.

[6] J. BRUNA and J. M. ORTEGA, Interpolation by holomorphic functions smooth to the boundary in the unit ball, *preprint.*

[7] D. BURNS and E. L. STOUT, Extending functions from submanifolds of the boundary, *Duke Math. J.* 43 (1976), 391-404.

[8] L. CARLESON, Sets of uniqueness for functions regular in the unit circle, *Acta Math.* 87 (1952), 325-345.

[9] J. del CASTILLO, On interpolation, peak and zero set on a weakly pseudoconvex domain, *Proceedings of 7th Spanish-Portuguese Conference on Math*, Part II, Publ. Sec. Mat. Univ. Autònoma Barcelona 21 (1980), 175-176.

[10] D. CATLIN, Global regularity of the $\bar\partial$-Neumann problem, *Complex Analysis of Several Variables,* Proc. Symp. Pure Math. 41, Am. Math. Soc., Providence, 1984, pp. 39-49.

[11] J. CHAUMAT and A.-M. CHOLLET, Ensembles pics pour $A^\infty(D)$, *Ann. Inst. Fourier (Grenoble)* 29 (3) (1979), 171-200.

[12] J. CHAUMAT and A.-M. CHOLLET, Caractérisation et propriétés des ensembles localement pics de $A^\infty(D)$, *Duke Math. J.* 47 (1980), 763-787.

[13] J. CHAUMAT and A.-M. CHOLLET, Ensembles pics pour $A^\infty(D)$ non globalement inclus dans une variété intégrale, *Math. Ann.* 258 (1982), 243-252.

[14] A.-M. CHOLLET, Ensembles de zéros à la frontière de fonctions analytiques dans des domaines strictement pseudo-convexes, *Ann. Inst. Fourier (Grenoble)* 26 (1)(1976), 51-80.

[15] A.-M. CHOLLET, Ensembles de zéros, ensembles pics pour A(D) et $A^\infty(D)$, *Complex Analysis (Québec)*, Progress in Math. 4, Birkhaüser, Boston, 1980, pp. 57-66.

[16] R. R. COIFMAN and G. WEISS, *Analyse Harmonique Non-Commutative sur Certains Espaces Homogènes*, Lect. Notes in Math. 242, Springer-Verlag, Berlin, 1971.

[17] J. P. D'ANGELO, Finite-type conditions for real hypersurfaces in $\mathbb{C}^n$, *preprint*

[18] A. M. DAVIE and B. K. ØKSENDAL, Peak interpolation sets for some algebras of analytic functions, *Pac. J. Math.* 41 (1972), 81-87.

[19] T. DUCHAMP and E. L. STOUT, Maximum modulus sets, *Ann. Inst. Fourier (Grenoble)* 31 (3) (1981), 37-69.

[20] F. FORELLI, Measures orthogonal to polydisc algebras, *J. Math. Mech* 17 (1968), 1073-1086.

[21] J. E. FORNAESS, Peak points on weakly pseudoconvex domains, *Math. Ann* 227 (1977), 173-175.

[22] J. E. FORNAESS and B. S. HENRIKSEN, Characterization of global peak sets for $A^\infty(D)$, *Math. Ann.* 259 (1982), 125-130.

[23] J. E. FORNAESS and S. G. KRANTZ, Continuously varying peaking functions, *Pac. J. Math.* 83 (1979), 341-347.

[24] J. E. FORNAESS and A. NAGEL, The Mergelyan property for weakly pseudoconvex domains, *Man. Math.* 22 (1977), 199-208.

[25] J. E. FORNAESS and N. ØVRELID, Finitely generated ideals in $A(\Omega)$, *Ann. Inst. Fourier (Grenoble)* 33 (1983), 77-86.

[26] I. GLICKSBERG, *Recent Results in Function Algebras*, Regional Conf. Series 11, Am. Math. Soc., Providence, 1972.

[27] J. GLOBEVNIK, Peak sets for polydisc algebras, *Mich. Math. J.* 29 (1982), 221-227.

[28] J. GLOBEVNIK, Norm preserving interpolation sets for polydisc algebras, *Math. Proc. Cam. Philos. Soc.* 91 (1982), 291-303.

[29] M. HAKIM and N. SIBONY, Quelques conditions pour l'existence de fonctions pics dans des domaines pseudoconvexes, *Duke Math. J.* 44 (1977), 399-406.

[30] M. HAKIM and N. SIBONY, Ensembles pics dans des domaines strictement pseudoconvexes, *Duke Math. J.* 45 (1978), 601-617.

[31] G. M. HENKIN and A. E. TUMANOV, Interpolation submanifolds of pseudoconvex manifolds, *Tr. Am. Math. Soc.* 115 (1980), 59-69.

[32] B. S. HENRIKSEN, A peak set of Hausdorff dimension 2n-1 for the algebra A(D) in the boundary of a domain D with C^∞-boundary in $\mathbb{C}^n$, *Math. Ann.* 259 (1982), 271-277.

[33] A. IORDAN, Peak sets in weakly pseudoconvex domains, *Math. Z.* 188 (1985), 171-188.

[34] A. IORDAN, Ensembles de module maximal dans des domaines pseudoconvexes, *C. R. Acad. Sci. Paris* Sér I 300 (1985), 655-656.

[35] A. IORDAN, Peak sets in pseudoconvex doamins with the (NP) property, *Math. Ann.* 272 (1985), 231-236.

[36] T. JIMBO, Peak sets on the boundary of a weakly pseudoconvex domain, *Math. Jap.* 29 (1984), 51-55.

[37] J.-M. LABONDE, *thesis*, Université de Paris-Sud, Centre d'Orsay, 1985.

[38] A. NAGEL, Smooth zero sets and interpolation sets for some algebras of holomorphic functions on strictly pseudoconvex domains, *Duke Math. J.* 43 (1976), 323-348.

[39] A. NAGEL, Cauchy transforms of measures and a characterization of smooth peak interpolation sets for the ball algebra, *Rocky Mt. J. Math.* 9 (1979), 299-305.

[40] A. NAGEL and W. RUDIN, Local boundary behavior of bounded holomorphic functions, *Can. J. Math.* 30 (1978), 583-592.

[41] A. V. NOELL, Properties of peak sets in weakly pseudoconvex domains in $\mathbb{C}^2$, *Math. Z.* 186 (1984), 99-116.

[42] A. V. NOELL, Interpolation in weakly pseudoconvex domains in $\mathbb{C}^2$, *Math. Ann.* 270 (1985), 339-348.

[43] A. V. NOELL, Differentiable peak-interpolation on bounded domains with smooth boundary, *Bull. Lond. Math. Soc.* 17 (1985), 134-136.

[44] A. V. NOELL, Peak points in boundaries not of finite type, *Pac. J. Math.* 123 (1986), 385-390.

[45] W. RUDIN, *Function Theory in Polydiscs*, W. A. Benjamin, New York, 1969.

[46] W. RUDIN, Peak-interpolation sets of class $\mathbb{C}^1$, *Pac. J. Math.* 75 (1978), 267-279.

[47] W. RUDIN, Holomorphic Lipschitz functions in balls, *Comment. Math. Helv.* 53 (1978), 143-147.

[48] R. SAERENS, Interpolation manifolds, *Ann. Sc. Norm. Sup. Pisa Cl. Sci.* IV Ser. 11 (1984), 177-211.

[49] R. SAERENS and E. L. STOUT, Differentiable interpolation on the polydisc, *Complex Variables* 2 (1984), 271-282.

[50] E. M. STEIN, Singular integrals and estimates for the Cauchy-Riemann equations, *Bull. Am. Math. Soc.* 79 (1973), 440-445.

[51] B. STENSØNES, Zero sets for A^∞ functions, *preprint*.

[52] E. L. STOUT, *The Theory of Uniform Algebras*, Tarrytown-on-Hudson, New York, 1971.

[53] E. L. STOUT, The dimension of peak-interpolation sets, *Proc. Am. Math. Soc.* 86 (1982), 413-416.

[54] B. A. TAYLOR and D. L. WILLIAMS, The peak sets of A^m, *Proc. Am. Math. Soc.* 24 (1970), 604-605.

[55] A. E. TUMANOV, A peak set for the disc algebra of metric dimension 2.5 in the three-dimensional unit sphere, *Math. USSR Izv.* 11 (1977), 353-359.

[56] R. E. VALSKII, On measures orthogonal to analytic functions in $\mathbb{C}^n$, *Soviet Math. Dokl.* 12 (1971), 808-812.

[57] N. T. VAROPOULOS, Ensembles pics et ensembles d'interpolation pour les algèbres uniformes, *C. R. Acad. Sci. Paris Sér. A* 272 (1971), 866-867.

[58] J. VERDERA, A remark on zero and peak sets on weakly pseudoconvex domains, *Bull. Lond. Math. Soc.* 16 (1984), 411-412.

[59] B. M. WEINSTOCK, Zero-sets of continuous holomorphic functions on the boundary of a strongly pseudoconvex domain, *J. Lond. Math. Soc.* 18 (1978), 484-488.

EXTENDABILITY OF HOLOMORPHIC FUNCTIONS

Berit Stensones*
Rutgers University
Mathematics Department
New Brunswick, New Jersey

Introduction

In this paper we are studying the extendability of holomorphic functions from one side of a real hypersurface in $\mathbb{C}^n$.

The proofs presented here is highly based on the maximum principle and the subaveraging property of plurisubharmonic functions.

Let $S: = \{p \in \mathbb{C}^n : r(p) = 0\}$ where r is a C^2 defining function. By the two sides S^+ and S^- of S we shall mean that $S^+: = \{p \in \mathbb{C}^n : r(p) > 0\}$ and $S^- := \{p \in \mathbb{C}^n : r(p) < 0\}$. If U is a neighborhood of a point $p \in S$, then $U^+: = S^+ \cap U$ and $U^-: = U \cap S^-$.

If U is small, then this makes sense, i.e. U^- and U^+ are open domains in $\mathbb{C}^n$.

We shall be interested in finding out whether holomorphic functions on U^+ or U^- has holomorphic extension over p.

Our main result is a generalization of a result by J.M. Trepreau [1] saying:

THEOREM 1 [1]:

Let S be a C^2-smooth hypersurface in C^n and let $p_o \in S$. Then $p_o \in \tilde{U}^+$ or $P_o \in \tilde{U}^-$ for every neighborhood U of p_o if and only if there is no germ of a complex hypersurface in S through p_O.

By $\tilde{D}$ we mean the envelope of holomorphy of a domain D.

In fact Trepreau is able to prove that if there is no germ of a complex hypersurface in S through p_o then the following holds:

Let V be a neighborhood of p_o in S and let U be a neighborhood of V in $\mathbb{C}^n$. Then there exists a neighborhood W of p_o in $\mathbb{C}^n$, W is independant of U

*The author partly supported by an N.S.F. grant.

(i.e. W is independent of how thick U is) such that $W \subset \tilde{U}^+$ or $W \subset \tilde{U}^-$.

Note that we do not know whether $W \subset \tilde{U}^+$ or $W \subset \tilde{U}^-$, we just know that W is contained in one of them.

What happens if there is a germ of a complex hypersurface in S through p_o? Can we by choosing V large enough still find a neigborhood W of p_o in $\mathbb{C}^n$ such that $\tilde{W} \subset U^-$ or $\tilde{W} \subset U^+$?

Of course if $\sum := \{p \in \mathbb{C}^n : h(p) = 0\} \subset S$ where h is a holomorphic function, then $\frac{1}{h}$ is holomorphic in U^+ and in U^-. Hence if $p_o \in \sum$, then p_o is not in $\tilde{U}^+$ or $\tilde{U}^-$.

On the other hand, if Δ is an analytic disc in S, S is a real hypersurface in $\mathbb{C}^2$, then we can prove that $\Delta \subset \tilde{U}^+$ or $\tilde{U}^-$ whenever U is a neighborhood of V and V is a neighborhood of $\bar{\Delta}$ in S.

THEOREM 2:

Let S be a C^2-smooth real hypersurface in $\mathbb{C}^n$ and let $p_o \in S$. Assume that there is a germ og a complex hypersurface $\sum$ in S through p_o. If there exists a point $p \in \sum \cap S$ such that p is not in the interior of $\sum \cap S$. Then $p_o \in \tilde{U}^+$ or $p_o \in \tilde{U}^-$ whenever U is a connected neighborhood of p and p_o.

By the interior of $\sum \cap S$ we shall mean the interior relative to $\sum$. By a fixed sized neighborhood we shall mean that it only depends on $U \cap S$ and not on the thickness of U (i.e. maxdistance from $p \in U$ to S).

MAIN SECTION

In this section we shall prove the main theorem and simple proof of Bochners theorem on extendability of C-R functions on compact real hypersurfaces in $\mathbb{C}^n$.

Before doing the actual proof we should make a couple of observations.

First we observe that if $q \in \overline{V}$ where V is the interior of $\sum \cap S$ where $\sum$ is a complex hyperplane (i.e. V is a germ of a complex hyperplane in S).

Then there is no other germ V_1 such that $q \in \overline{V}_1 \subset S$.

The reason for this is that the complex tangentplane $T_{\mathbb{C}} S$ of S and the tangentplane TV of V are equal all over V. Hence V is locally an integral manifold for the complex structure on S and it is of the same dimension as $T_{\mathbb{C}} S$. (i.e. V is of maximal dimension.) From the uniqueness of integral curves γ of a vectorfield n with boundary conditions $Z \in \gamma$ we can deduce that if $Z \in \overline{V}_1$, then $V_1 \cap V$ is open. Hence $V_1 = V$.

Next, we observe that S is C^2-smooth so it will not turn around and touch itself. So if W_1 and W_2 are two neighborhoods of points p_1 and p_2 in S such that function f holomorphic in say U^- extends holomorphically to f_1 on $U^- \cup W_1 \setminus U^-$ and f_2 on $U^- \cup W_2 \setminus U^-$. Then we can by shrinking W_1 and W_2 assume that they are balls. Hence $W_1 \cap W_2$ are connected. If we let W_1 and W_2 be balls with centers p_1 and p_2, then $W_1 \cap W_2 \neq \emptyset$ will imply that $W_1 \cap W_2 \cap U^- \neq \emptyset$. All this will imply that $f_1 = f_2$ on $W_1 \cap W_2$, so we have a holomorphic extension of f to $U^- \cup ((W_1 \cup W_2) \setminus U^-)$.

Now to the main part of the proof of Theorem 2. From the first observation and the conditions on p given in Theorem 2 we know that there is no germ of a complex hyperplane in S through p.

Choose a curve $\gamma : [0,1] \to \sum \cap S$ such that $\gamma(0) = p_0$ and $\gamma(1) = p$. Furthermore we want $\gamma(\langle 0,1 \rangle)$ to be contained in the interior of $\sum \cap S$ and $\gamma([0,1])$ should have finite length.

From Trepreau's theorem we know that there exist a neighborhood W_1 of $\gamma(1) = p$ such that all holomorphic functions on U^- or U^+ say U^- extends to $U^- \cup (W_1 \setminus U^-)$. Note that we do not know from which side $(U^+$ or $U^-)$ the functions extend. We also know that the size of W_1 only depends on the size of $U \cap S$ and of the shape of S.

If $p_o \in W_1 \cap S$, then we are done. If not, let p_1 be a point in $\gamma([0,1\rangle)$ closer to p_o than p is. Let $\gamma(p_1,p)$ be the piece of γ between p and p_1.

Now we lift S off Σ along $\gamma(p_1,p)$ in the following may:

Let S_1 be a C^2-smooth surface such that

i) $S_1 \setminus W_1 = S \setminus W_1$

ii) $(\gamma(p_1,p) \cap S_1) \cap S = p_1$

i.e. $\gamma(p_1,p)\setminus\{p_1\}$ is not contained in S_1.

iii) There exists a neighborhood V of $\gamma \setminus \overline{\gamma(p_1,p)}$ such that $V \cap S_1 = V \cap S$.

iv) $S_1 \setminus S \subset W_1 \setminus \overline{U^-}$.

In other words we lift S off $\gamma(p_1,p)$ into $W_1 \setminus U^-$, makes sure that $p_1 \in S$ and leaves the rest unchanged.

Condition i,ii and iii together with the first remark in this section ensures that there is no germ of a complex hypersurface in S_1 through p_1. The size of $W_1 \setminus \overline{U^-}$ determines how much we can lift S off $\gamma(p_1,p)$.

If $S_1 := \{p \in \mathbb{C}^n : r_1(p) = 0\}$ where r_1 is a C^2-defining function and $U_1^- = U \cap \{p \in \mathbb{C}^n : r_1(p) < 0\}$, then $U^- \subset U_1^-$ and all $f \in H(U^-)$ extends holomorphically to U_1^-, i.e., $H(U-) = H(U_1^-)$. This last statement follows from i and iv.

Now we can find a neighborhood W_2 of p_1 such that all functions $f \in H(U_1^-)$ extends holomorphically to $U_1^- \cup W_2$. Hence all $g \in H(U^-)$ extends to a holomorphic function in $U_1^- \cup W_2$. This is obtained by applying Theorem 1 and observing that $\Sigma \subset S_1^+$ locally, hence not all functions holomorphic in $V \cap S_1^+$ extends over p_1 if V is a small neighborhood of p_1.

Choose a point $p_2 \subset W_2 \cap \gamma$ closer to p_o than p_1 and let $\gamma(p_1,p_2)$ be the part of γ in W_2 from z_2 to z_3.

Now we lift S_1 off Σ along $\gamma(p_1,p_2)$ and get a C^2-smooth hypersurface S_2 where:

i) $S_2 \setminus W_2 = S_1 \setminus W_2$

ii) S_2 does not contain $\gamma(p_1,p_2)$ except the point p_2.

iii) $p_2 \in S_2$

iv) There exists a neighborhood V of $\gamma \setminus \overline{\gamma(p_1,p_2)}$ such that $V \cap S_2 \quad W_2 \cap U_1^-$

v) $S_2 \setminus S_2 \subset W_2 \setminus U_1^-$.

As above we obtain a neighborhood W_3 of p_2 such that all functions $g \in H(U^-)$ extends to $(U \cap S_2^-) \cup W_3$.

By S_2^- we shall mean $\{p: r_2(p) < 0\}$ where r_2 is a C^2-defining function for S_2 chosen so that $S_2^- \supset S_1^- \supset S^-$.

We can keep doing this, i.e., lifting S off $\gamma(p_{i-1},p_i)$ until p_0 is contained in W_{i+1}. We can prove that there exists a $r > o$ such that $B(p_i,r) \cap \sum \subset W_{i+1}$ for each i the following way:

Locally near p_i we may consider S to be a graph. In other words $S = \{(z^1,w) : v = \rho\,(z^1,u)\}$ here $z^1 = (z_1,\ldots,z_{n-1})$ and $w = u + iv$.

Say U^- is locally $\{v < \rho(z^1,u)\}$. Then $\widetilde{U^-}$ is locally schlicht, i.e. $\widetilde{(U \cap B)}$ is a domain in $\mathbb{C}^n$ if B is a small neighborhood of pi. Let us write $\widetilde{U^-}$ instead of $\widetilde{(U \cap B)}$.

Let $\mathbb{B}_\sum(p_i,R)$ be a ball of radius R and center p_i in $\sum$. If $i > 1$ then such a ball exist and $R > o$. In other words $\mathbb{B}_\sum$ is a ball inside $\sum$ in the sense of $\sum$'s local coordinates.

If $q \in \mathbb{B}_\sum(p_i,\frac{R}{4})$, then $\mathbb{B}_\sum(p_i,\frac{R}{4}) \subset \mathbb{B}_\sum(q,\frac{R}{2}) \subset \mathbb{B}_\sum(p_i,R)$.

Let $t > 0$ be small, then $\sum_t = \{p - t(0,i) \in U^-\}$ and $\sum_t$ is still a complex hyperplane. If $\rho = -\log \operatorname{dist}(\cdot, \partial\widetilde{U^-})$, then ρ is a plurisubharmonic on $U^- \subset \widetilde{U^-}$ since $\widetilde{U^-}$ is Stein. Hence $\rho \mid \sum_t$ is plurisubharmonic for all $t > 0$.

Let $q \in \mathbb{B}_\sum(p_i,\frac{R}{4})$ we want to prove that this implies that $q \in \widetilde{U^-}$. If we choose $r > 0$ small enough then $\mathbb{B}(p_i,r) \cap \sum \subset \mathbb{B}_\sum(p_i,\frac{R}{4})$ and we can let W_{i+1} contain $\mathbb{B}_\sum(p_i,\frac{R}{4})$.

Assume $q \in \mathbb{B}_\sum(p_i,\frac{R}{4})$ but $q \in \widetilde{U^-}$ and let $\mathbb{B}_{\sum t}(q,\frac{R}{2}) = \{p - t(0,i): p \in \mathbb{B}_\sum(q,\frac{R}{2})\}$ call this set B_t. Since $\mathbb{B}_\sum(p_i,\frac{R}{4}) \subset \mathbb{B}_\sum(q,\frac{R}{2})$ and

$\gamma(p_{i-1}, p_i) \cap \mathbb{B}_\Sigma(p_i, \frac{R}{4}) \neq \emptyset$, then there exists a set A_t $\mathbb{B}_{\Sigma t}(q, \frac{R}{2})$ such that $\mathrm{dist}(A_t, \partial\tilde{U}^-) > b > 0$ for all t and volume $(A_t) > a > 0$ for all t.

By using the subaveraging principle for plurisubharmonic functions we get:

$$\rho(q) < \frac{1}{\mathrm{vol}(\mathbb{B}_t)} \Big(\int_{\mathbb{B}_t \setminus A_t} \rho \, d\lambda + (-\log a)\, \mathrm{vol}(A_t)\Big)$$

Assume $\mathrm{vol}(\mathbb{B}_t) = 1$ and observe that if $p \in S$, then locally $\mathrm{dist}(p - t(0,i), S) \sim t$. Hence there exists a constant $c > o$ such that

$$\max \{\rho : \mathbb{B}_t\} < -\log(C\, t) = -\log t + C'.$$

Furthermore if $q \notin \tilde{U}^-$, then $\mathrm{dist}(q - t(0,i), S) \approx \mathrm{dist}(q - t(0,i), \partial\tilde{U}^-)$. Hence there exists a constant C_1 such that $\rho(q) > -\log t - C_1$.

All this gives that $-\log t < (1-b)(-\log t) - (\log a)b + C_2$. If we let t go to zero we obtain a contradiction.

The reasion why p_0 is reached at some point is the fact that $\gamma(z_i, z_{i+1})$ can be choosen to have a finite length.

<u>COROLLARY</u>:

Let M be a complex manifold without bounded complex hypersurfaces. If D is a bounded domain with a C^2-smooth boundary. Then there exist a domain Ω and a compact $K \subset\subset \Omega$ such that $D \subset \Omega$ and all C-R functions on ∂D extends to a holomorphic function in $\Omega \setminus K$.

<u>PROOF</u>:

First we observe that there is no bounded domain in M containing a complex hypersurface Σ in its boundary. Hence, if there is a germ $\mathrm{int}(\Sigma \cap \partial D)$ [this is the interior with respect to Σ] of a complex hypersurface Σ in ∂D, then there is a point $p \in \Sigma\ \partial D$ so that p is not contained in $\mathrm{int}\Sigma \cap \partial D$ is compact.

This implies that whenever $z \in \partial D$, then there is a fixed sized neighborhood $\tilde{W}(z)$ of z such that C-R functions on ∂D does have a holomorphic extension to $W(z)$ where $W(z) := \{\tilde{W}(z) \setminus \overline{D}$ or $\tilde{W}(z) \cap D\}$ depending on to which side they extend. Since ∂D is compact, then we may assume that the size of $W(z)$ does have a minimum.

Let $R(D) := \bigcup_{z\in\partial D} W(z)$, then all C-R functions on ∂D does have a holomorphic extension to $R(D)$.

Define $\Omega := \mathrm{int}(R(D)\cup\bar{D})$ and $K := \bar{D}\setminus R(D)$. Then $D\subset\Omega$ and $R(D) = \Omega\setminus K$.

REFERENCES

1. J.M. Trépreau: "Sur le prolongement holomorphe des functions C-R definies sur une hypersurface reelle de classe C^2 dans $\mathbb{C}^n$." Invent. Math. 83, 583-592 (1986).